The 21st Century Meeting and Event Technologies

Powerful Tools for Better Planning, Marketing, and Evaluation

The 21st Century Meeting and Event Technologies

Powerful Tools for Better Planning, Marketing, and Evaluation

Seungwon "Shawn" Lee, PhD
Dessislava Boshnakova, PhD
Joe Goldblatt, EdD

APPLE
ACADEMIC
PRESS

Apple Academic Press Inc.	Apple Academic Press Inc.
3333 Mistwell Crescent	9 Spinnaker Way
Oakville, ON L6L 0A2	Waretown, NJ 08758
Canada	USA

©2017 by Apple Academic Press, Inc.

Exclusive worldwide distribution by CRC Press, a member of Taylor & Francis Group

No claim to original U.S. Government works

Printed in the United States of America on acid-free paper

International Standard Book Number-13: 978-1-77188-023-7 (Hardcover)

International Standard Book Number-13: 978-1-4822-5184-5 (eBook)

Library and Archives Canada Cataloguing in Publication

Lee, Seungwon, author
The 21st century meeting and event technologies : powerful
tools for better planning, marketing, and evaluation / Seungwon
"Shawn" Lee, PhD, Dessislava Boshnakova, PhD, Joe Goldblatt, EdD, FRSA.

Includes bibliographical references and index.
Issued in print and electronic formats.
ISBN 978-1-77188-023-7 (hardcover).--ISBN 978-1-4822-5184-5 (pdf)
1. Special events--Technological innovations. 2. Special events--Management. 3. Special events--Planning. I. Goldblatt, Joe Jeff, 1952-, author II. Boshnakova, Dessislava, author III. Title. IV. Title: Twenty-first century meeting and event technologies.

GT3405.L44 2015	394.2	C2015-906540-2	C 2 0 1 5 -
906541-0			

Library of Congress Cataloging-in-Publication Data

Names: Lee, Seungwon, author.
Title: The 21st century meeting and event technologies : powerful tools for better planning, marketing and evaluation / Seungwon Lee, Dessislava Boshnakova, Joe Goldblatt.
Other titles: Twenty-first century meeting and event technologies
Description: Mistwell Crescent, Oakville, ON ; Waretown, NJ : Apple Academic Press, [2015] | Includes bibliographical references and index.
Identifiers: LCCN 2015036706 | ISBN 9781771880237 (alk. paper)
Subjects: LCSH: Special events--Technological innovations. | Special events--Management. | Special events--Planning.
Classification: LCC GT3405 .L44 2015 | DDC 394.2--dc23
LC record available at http://lccn.loc.gov/2015036706

Apple Academic Press also publishes its books in a variety of electronic formats. Some content that appears in print may not be available in electronic format. For information about Apple Academic Press products, visit our website at **www.appleacademicpress.com** and the CRC Press website at **www.crcpress.com**

Dedication

This book is dedicated with appreciation and love to my family who has been the best supporters from my day one in this world. My mother, Young-Ae Kang, and father, Dong-Bok Lee, who was also an educator, have supported my education and, most of all, taught me the importance of working hard with respect to others. My special heartfelt words of thanks to my wife, Young-seon Nam, my daughter, Jiyoon, and my son, Brandon, who have been always next to me with warm and endless love. In addition, I want to express my sincere appreciation to my two academic advisors from my doctoral and master degree studies. Dr. Mahmood Khan from Virginia Tech and Dr. Joe Goldblatt, who is also co-author of this book, have guided me throughout every steps of my academic journey in both sunny days and rainy days. They deserve far more credit than I do for the publication of this book. I consider myself to be blessed beyond description to dedicate this book to all of the above-mentioned people.

— Seungwon "Shawn" Lee, PhD,
George Mason University, Fairfax, VA, USA

I want to dedicate this book to my beloved family. They support me in the journey and have been there to help me at any time. Warm thanks to my husband, Edi, to my son, Andi, to my daughter, Dimana and to my mother, Rumi.

Two more persons have a big impact on my passion for special events and communication. In the year 1998 Professor Vladimir

Mihaylov invited me to become part of the Department of Mass Communication at the New Bulgarian University. He has opened for me the door to the great world of teaching, and I am so thankful for this. At the beginning of the 21st century when translating from English to Bulgarian for a book with Dr. Joe Goldblatt, I discovered not only an inspiring teacher, but also a great friend. I would not be part of this book if those two professors had not crossed my world.

At the end I want to dedicate this book to all students who will find inspiration in it for their future careers.

—Dessislava Boshnakova, PhD,
New Bulgarian University, Sofia, Bulgaria

Dr. Seungwon "Shawn" Lee and Dr. Dessislava Boshnakova are not only my co-authors, they are also my former students. Therefore, I dedicate this book to the next generation of meeting and event technology students who will hopefully emulate the remarkable achievements of Seungwon and Dessislava and indeed the thousands of former students that I have had the privilege to introduce to the boundless world of planned events.

My first book was written by myself nearly 30 years ago and has served as a primer and introductory text for event planning for many years. Therefore, I am exceedingly pleased and proud that this book is the work of three pairs of hands that include my brilliant students and now esteemed colleagues, Seungwon and Dessislava.

—Joe Goldblatt, EdD, FRSA,
Queen Margaret University, Edinburgh, Scotland

Contents

Foreword

These are exciting times. Half the scientists that have ever lived are alive today and, thanks to the Internet, they are far better connected than at any other time in history. Consequently, we are going through a renaissance of discovery in science, medicine, and technology, which will affect all of our lives and transform most industries, including meetings and events.

The rate of technology change is accelerating with thousands of ideas, apps, and innovations bubbling up to help meeting planners, exhibitors, venues, and other meeting participants to do their jobs better. This digital revolution is profoundly changing events. It is streamlining many logistics tasks, improving collaboration, driving down costs, and greatly enriching the attendee experience.

The challenge for many meeting professionals, however, is how does one keep up? This is where *The 21st Century Meeting and Event Technologies* can help. The contents cover a broad range of events technology issues with thoughtful insight. This book fills a substantial need for meeting professionals to make sense of the rapidly changing plethora of choices available.

Starting with the history of event technology, this important work covers the strategic use of technology. It progresses through an analysis of several technology product categories, It covers many important marketing issues and then looks to the future. It is full of case studies, additional video content links, and more.

I heartily commend the efforts of Drs. Lee, Boshnakova, and Goldbatt in putting together this important academic work that will be highly useful for both students and the meeting professionals wishing to advance their career.

If you are looking for well-rounded coverage of a wide range of events technology issues with lots of specific ideas for adoption, selection and implementation, I highly recommend this book.

Corbin Ball, CMP, CSP, DES
Meetings Technology Analyst
Corbin Ball & Co.

Preface

This new and first ever book in the field of meeting and event management is emblematic of the rapid changes occurring throughout the world with regard to research, communications, planning, coordination and evaluative technologies. This book could not have been written twenty years ago because at that time these technologies simply did not exist.

During the past two decades the three authors responsible for developing this textbook have had the opportunity to witness from a front row seat the sweeping technological changes that have changed the meetings and events landscape forevermore. These changes encompass increased speed in decision making, advanced access to information and data, and the ability to automate many functions within meeting and event management that at one time would have required substantial human labor.

WHY WE WROTE THIS BOOK FOR YOU

Despite the wide differences in age and culture and experience, the authors share a strong interest in and commitment to developing knowledge about the future of meeting and event technology. The reason why we have invested nearly two years of our life to research and writing this book is actually quite simple. We believe that for you to earn your first job and to continually be promoted in the meetings and events field, it is essential that you master the fundamentals and as well as some advanced skills in technology.

Only three decades ago technology for meetings and events was the equivalent of an electronic typewriter. In the intervening thirty years technology has become all pervasive in every function of the meeting and event planning job description. Therefore, for you to get in the game, stay in the game and grow financially in the future, you must stay ahead of the game that is rapidly developing around technology.

In order for you to be successful in obtaining an entry-level job, you will need to master basic technology skills in data entry, research, analysis, and report writing, as well as be able to enumerate the financial plan for your meeting or event through the development and management of a budget. In addition to these basic skills, you will need specialized training in computer-assisted design and drawing to create floor plans and diagrams as well as the ability to develop applications for mobile devices that will be used by your attendees to monitor the programs you create.

Perhaps most importantly, in order for you to progress in your career, you will need to be able to use technology to carefully measure and evaluate the return on investment (ROI), return on marketing investment (ROMI) and return on objective (ROO) for every meeting and event. Therefore, mastering programs such as SurveyMonkey and MeetingMetrics may be in the future just the tip of the iceberg of what you will need to know to demonstrate to your superiors that your meeting and event is successful and to promote continuous improvement.

While social media is perhaps the most dramatic technological change in meeting and event marketing, it is also one of the most difficult to measure and evaluate. Therefore, in the future not only will you be programming applications that are already commercially available, you may indeed be working with future software designers to create custom apps for your meeting and event so that you may more carefully measure your outcomes.

The term 'event' is derived from the Latin term 'e' meaning out and 'venire' meaning 'come'. Therefore, every time you create a meeting or event, you are in fact designing an outcome. Meeting and event technology will allow you to better achieve these outcomes with much great efficiency, speed, consistent overall quality and improved success for your organization.

STRUCTURE OF THE BOOK

This book is structured in four parts. Part One explores the traditions and upward trajectory of meeting and event technology. The chapters in this first part examine how technology has developed and where it may be going in terms of future development.

Chapter 1 examines in detail the 75-year history of meeting and event technology and discusses how the tools for research, design, planning, coordination and evaluation have dramatically changed through miniaturization of design and speed of broadband.

Chapter 2 discusses how to use the right technology for the right purpose. In this chapter you will learn how to conduct internal research to see where your technology gaps are within your organization and then how to identify the proper technologies to close these gaps in the future.

Chapter 3 explores the concept of search engine optimization (SEO), which is so important if you want the content about your event to be accessible and easy to find online. SEO strategies will help you to position your meeting and event at the very top of Google and other search engine rankings.

Chapter 4 discusses the importance of using search engines to conduct your preliminary destination and venue research. In this chapter you will learn how to not only search for venues but also how to use computer software to design your meeting and event space.

Part 2 follows on from chapter four by further introducing you to practical technological skills in design, planning and evaluation.

Chapter 5 provides additional techniques for designing and decorating your meeting and event space to achieve the outcomes you desire from your attendees.

Chapter 6 provides you with the critical tools for developing the meeting and event budget and monitoring your performance in real time through advanced technological tools.

Chapter 7 explores the new frontier of virtual events. In this chapter you will learn about synchronous as well as asynchronous communication and learning opportunities to grow the educational opportunities for your stakeholders.

Chapter 8 describes how technology may be used to create and execute a 360 degree evaluation of your meeting and event. It also discusses how you may use a mixed evaluation (pre and post, on site) to produce more high quality information for you for future decision making.

Part 3 enters the extremely important meeting and event domain of marketing. This part demonstrates how you may use the multi-expanding platforms in social media to increase attendance and engagement your meeting and event.

Chapter 9 demonstrates how to develop, design, and manage wikis, websites, audio and video blogs and podcasts to further engage your participants before, during and after your meeting or event.

Chapter 10 provides you with tools and techniques you will need to conquer and expand the new frontiers of social media. Not only will you learn about the basic social media platforms used in meetings and events and business in general, you will also learn how to integrate these marketing channels to improve the return on your marketing investment.

Chapter 11 enables you to go mobile with your communications through a thorough introduction to numerous new applications and the challenges and opportunities you will face in a unplugged meeting environment.

Chapter 12 focuses on how to encourage, grow, monitor, manage and develop user-generated content to promote deeper engagement among your participants.

Chapter 13 carefully and thoughtfully introduces you to registration and transaction systems to ensure that you make it easy for your participants to join your meeting and event and that you are able in real time to monitor your progress.

Chapter 14 concludes Part 3 with one of the most recent developments in meeting and event technology as it explores the many opportunities available to you through crowdsourcing potential attendees.

The final part of the book allows you to pick up the crystal ball crafted by the authors to look at the potential future developments in meeting and event technology. Part Four also includes over 50 case studies from actual meeting and event technology situations and the best practices the organizers employed to improve the outcome through technological innovation. This final part concludes with a wide range of additional resources to assist you in your further study.

Chapter 15 previews the current and future developments in holograms, robotics and other technological innovations that may reduce the carbon footprint for your event while creating the excitement you need to differentiate your program from competitors.

In some chapters you will find brief anecdotal case studies called "Screenshots" to further illustrate the information you are reading. These case studies are based upon real-world examples that were part of the development of technology for the modern meetings and events instrument during the past half century.

HOW TO BEST USE THIS BOOK

Each chapter includes carefully selected and prepared endnotes to encourage you to discuss, problem solve and direct you to additional resources. Make certain you take time to complete the questions and discussion at the end of each chapter and then refer to the additional resources you may need for further study.

The authors have also video-captured interviews of many of the key individuals featured in this text. You are strongly encouraged to visit the video links available from the publisher (http://www.appleacademic-press.com/video_interviews/9781771880237) to help you better understand the principles and personalities associated with the individuals who are helping lead the development of these new meeting and event technologies.

Finally, Part 4 features numerous case studies and additional resources. When you read the case study, pause and reflect before you read how the meeting and event organizer used technology to turn a challenge into a gainful opportunity. Compare your observations with those of others to practice your problem-solving skills. With regard to the additional resources, set a schedule to master one new technology platform per month. At the end of the year you will be able to add ten new skills to your growing resume!

THE TROUBLE WITH THE FUTURE

The French poet Paul Valery (1871–1945) famously wrote that "The trouble with our times is that the future is not what it used to be." The same is true for meeting and event technology. Whilst researching and writing this book, several knowledgeable technology experts told us that before we entered the words in the computer, the concepts would be outdated. We do not agree. In fact, we have found that, in fact, the past 75 years of meeting and event technological development have proven that from the manual typewriter to the smartphone and even Google Glass and Samsung Smart Watch, the needs of meeting and event professionals and their participants have remained very much the same. The need to conduct research to avoid future error and improve performance, the need to communicate, the need to design and plan with creativity and precision, and finally and perhaps most importantly, the need to measure and evaluate performance are even more important today than they were nearly one century ago. What has changed is the speed and intensity of these increasingly challenging demands.

According to Dan Berger, CEO of the Washington, DC-based event technology company Social Tables,

> At its core use case, meeting and event technology eliminates inefficiencies, maximizes collaboration, generates actionable reports, and engages participants. This, however, is only the tip of the iceberg. Advanced meeting and event technology should leverage data, identify patterns, and apply thought leadership to enhance meetings and events.
>
> For example, the subject of meeting design and effectiveness has been studied for years, and yet planners continue to use ineffective room sets (e.g. theatre for continuing education sessions or traditional U shapes for boardroom style meetings). This is mostly because hotels have always been in control of the diagram and—quite frankly–aren't incentivized to ensure the success of a meeting.
>
> Today's diagramming solutions (e.g. Social Tables) are finally liberating the hotel from the responsibility of owning the floor plan. They allow planners to create the room set that aligns with their meetings' objectives. But this is only the beginning.
>
> Smart diagramming solutions have a recommendation engine to make intelligent, research-backed meeting design suggestions to the planners using them in order to further drive their ROO (return on objectives).
>
> Taking this concept one step further, future iterations of such software can leverage machine learning to drive meeting effectiveness by comparing survey results to room setups in order to identify the most effective meeting design concepts. (Don Berger, personal communication)

In 2014, Berger's three-year-old meeting and event technology business received $8 million in Series A funding to promote further growth. When asked what has helped fuel Social Tables' phenomenal growth, he stated that "There are two things I'm very passionate about, and I believe they are actually one and the same: meeting design and meaningful connections. So many meetings and events are poorly designed (from the content they include to the way they are set-up), and so many people leave these experiences feeling unfulfilled. This is the problem we are working on solving at Social Tables."

Therefore, we hope that you will use this and other tools to advance your career and perhaps even develop tools yourself to improve the overall profession of meeting and events in the near future.

That is why we wrote this book: Not just for your needs today or tomorrow but to provide you with a basic reference that might serve as the foundation for your thinking about how to invent the technologies that are still unknown but will be needed in the near future. Good luck as you re-think and re-invent the future of meetings and events through these and many future emerging technologies.

Seungwon "Shawn" Lee, PhD, Associate Professor
George Mason University Fairfax, Virginia, USA

Dessislava Boshkanova, PhD, Professor
New Bulgaria University Sofia, Bulgaria

Joe Goldblatt, EdD, Professor
Queen Margaret University Edinburgh, Scotland

Ackowledgment

The publication of *The 21st Century Meeting and Event Technologies: Powerful Tools for Better Planning, Marketing and Evaluation* was supported by many people and organizations. We offer our sincere appreciation to those who helped our publication of this book.

Writing this book has been the dream come true and I have never had so much fun in my life. I could interview people who have advanced and provide visions to the meeting and event technology industry.

For our book, we interviewed key meeting and event technology industry leaders and pioneers who were willing to share their visions and insights with the readers. Therefore, we are very grateful to Mr. Corbin Ball, Mrs. Maria Gergova, Mr. David Meerman Scott, Mr. Plamen Russev, Ms. Michelle Russell, Ms. Betsy Bair, Mr. Dan Berger, Ms. Barbara Palmer, Mr. Colin Loveday and Mr. Bob Vaez.

We also appreciate George Mason University Office of Instructional Technology for its support in editing the video interviews that accompany each chapter.

We also thank Corbin Ball, a highly respected and well-known meetings technology speaker, consultant, and writer, for writing a thoughtful foreword to this book.

In addition, we also wish to acknowledge Apple Academic Press and its president Mr. Ashish Kumar, who has provided us an opportunity to publish this first of its kind, dedicated event and meeting technology textbook for students and industry practitioners. Their belief of this is the right topic and right time to publish has made this happening.

Ackowledgment

The publication of the 21st Century Meeting and Event Technologies: Powerful Tools for Better Planning, Marketing and Application was supported by many people and organizations. We offer our sincere appreciation to those who helped our publication of this book.

Writing this book has been the dream come true and I have never had so much fun in my life. I could interview people who have advanced and provide visions to the meeting and event technology industry.

For our book, we interviewed key meeting and event technology industry leaders and pioneers who were willing to share their visions and insights with the readers therefore we are so grateful to Mr. Corbin Ball, Mrs. Shana Cerpot, Mr. David Merriman Scott, Mr. Plamen Ivanov, Ms. Michelle Russell, Mr. Tracy Hall, Mr. Don Rergm, Mr. Barbara Palmer, Mr. Colin Lowndy and Mr. Bob Vees.

We also thank the George Mason University Office of Instructional Technology for its support in editing the video interviews that accompany each chapter.

We also thank Corbin Ball, a highly respected and well-known meeting technology speaker, consultant, and writer, for writing a thoughtful foreword to this book.

In addition, we also wish to acknowledge Apple Academic Press and its president Mr. Ashish Kumar, who has provided us an opportunity to publish the first of its kind, dedicated event and meeting technology textbook for students and industry practitioners. Their belief of the right topic and right time to publish has made this happening.

About the Authors

The authors of this book are representative of the changing landscape of members of the meetings and events industry. The three authors of this book represent three home countries (South Korea, Bulgaria, and the United States) and have a combined expertise in meetings and events of nearly 75 years.

The lead author, Dr. Seungwon (Shawn) Lee, from Fairfax, Virginia in the USA, and originally from Seoul, Korea, earned his masters and doctorate degrees in the United States. Dr. Lee worked as a computer programmer, an association meeting planner, and a director of special events for international organizations in Korea and USA before starting his academic career. He is now an associate professor who leads the meetings/events management and technology program at George Mason University.

The second author, Dr. Dessislava (Dessi) Boshnakova, from Sofia, Bulgaria, is the founder and owner of her PR agency. She is also a professor of public relations and events for the first private university in Bulgaria, New Bulgarian University. She met the third author when she translated one of his textbooks into Bulgarian.

The third and final author is Professor Joe Goldblatt, from Edinburgh, Scotland. Joe is the author, co-author, and editor of 33 books in the field of planned events. He recognizes that he is older than the lead and second author and, therefore, has less experienced with regard to developing technologies because of his late introduction to the Internet.

While Shawn and Dessi may be digital natives, Joe describes himself as a curious and enthusiastic digital immigrant. The blend of the authors' cultural backgrounds, life experiences, and meeting and event technology experience and skill have greatly contributed to the final outcome of this book.

Introduction

Welcome to the 21ˢᵗ Century and Beyond: The World of Meeting and Event Technology

The way people conduct business and communicate in the meeting and event industry has drastically changed over the last decade. Meeting and event technology used to be elemental to the event management and marketing. However, increased use of technology to research, plan, market, deliver, and evaluate events is very significant in modern meeting and event planning. Modern meeting and event management utilizes a model of integrated database systems to simplify and ensure greater accuracy for event management and marketing works. That

---

SCREENSHOT

Five steps to acquire and maintain the right technology (Goldblatt 2013):

1. Identify the technology needs within your organization.

2. Review and select appropriate technology.

3. Establish a schedule for implementation.

4. Provide adequate training for all personnel.

5. Review your needs systematically on a quarterly basis and adapt/adopt new technologies as required.

---

means that event research, development, marketing, coordination and evaluation have integrated into one interconnected complex but very efficient system to automate and expedite event management activities.

The meeting and event industry is a global business. According to the World Economic Forum (2008), it is important for event organizations to have core competencies (it has listed a total of twelve pillars) to effectively compete in the global marketplace. One of the twelve pillars of competitiveness within the global economy is "technological readiness." Such effort to lead in technology goes long way back somewhat sparingly. A concept of "eventology" was first explored 20 years earlier by the Institute for Eventology in Japan. It incorporates previous studies in sociology, anthropology, psychology, business, communications, theology, and "technology." Goldblatt (2002) has developed his "four-pillar approach: foundation for success," which suggests time, finances, human resources, and technology are what one needs to be a successful event professional.

While many technologies become relatively easier to operate, understanding the fundamentals of technologies and developing core competency in selecting the proper tools for meeting and event management is now critical to be a successful modern event planner than ever. Today's and tomorrow's event planners must remain current with all emerging technologies to create greater efficiency and increase overall quality of events.

TRENDS IN MEETING AND EVENT TECHNOLOGIES

Fast development of hardware along with affordable prices of tools, such as smartphones, tablets, PCs and devices using wireless fidelity (Wi-Fi) and radio frequency identification (RFID) lead the technology phenomenon. Wi-Fi is widely used in most tourism and hospitality businesses (hotels, convention centers, restaurants) as an alternative to a wired local area network (LAN). Many airports, hotels, and fast-food facilities offer free public access to Wi-Fi networks. In addition, technological infrastructure, such as fast and secured broadband equipped by latest fiber optic cables and hardware, has enabled many applications. They include social media, virtual meeting technology, three-dimensional event venue tours, cloud-based event floor design, and sharing, and so on. Social media, including Facebook, Twitter, LinkedIn, and Instagram are transforming the way in which event attendees communicate before, during and after events and meetings.

Changing demographics in the event and meeting industry is one of key factors that influence future directions of the meeting and event industry. It is well recognized that multiple generations are now attending meetings and events. They include pre-boomers, baby boomers, and Gens X, Y, and Z. Each generation has a very diverse way of doing business and communicating. Their wants and needs are also very diverse. Therefore, modern event professionals must be flexible and accommodate the various types of customers in terms of contents and delivery channels in their events. Plus, there is one more growing factor—the event and meeting industry is increasingly diverse in cultural background as meeting and event attendees are coming from all around the world.

The rapid technological development in the field of meetings and events will accelerate even more. Technology can create many new positions for modern event planners while bringing technological challenges to many event professionals at the same time. It is in this realm that modern event planners and those who are about to get into the field of event management need to improve their technological skills along with the fundamental skills of traditional event management. Fields related to the meeting and event industry, such as the tourism and hospitality industry, have experienced both high-tech and high-touch, which will be key to successful meeting and event management.

This textbook will help both students and entry-level meeting/event industry professionals better understand this critical topic: technology. Therefore, this book aims to provide you with the technological tools and resources you will need to thrive in the global meetings and events marketplace. This book seeks to serve as the first of hopefully many future volumes to come that will explore and expand the opportunities for meeting and event technologists to disseminate new information to promote greater access to education and encourage collaboration across many markets throughout the world.

REFERENCES

Goldblatt, J. (2002). *Special Events: Best Practices in Modern Event Management* 2nd ed., New York, NY: John Wiley & Sons, Inc.

Goldblatt, J. (2013). *Special Events: Creating and Sustaining a New World for Celebration.* 7th ed.. New York, NY: John Wiley & Sons, Inc.

World Economic Forum, (2008). Global Competitiveness Report, 2008-2009, viewed 1 October 2014: http://www.weforum.org/reports/global-competitiveness-report-2008-2009

PART I

The Traditions and Trajectory of Meeting and Event Technology

CHAPTER 1

The History of Meeting and Event Technology

> "Here's to the crazy ones. The misfits, the rebels, and the trouble-makers. The round pegs in the square holes. The ones who see things differently . . . they change things.
>
> They push the human race forward.
>
> And while some may see them as the crazy ones, we see genius."
>
> —Steve Jobs, Founder of Apple™ (1955–2011)

LEARNING OUTCOMES

As a result of reading this chapter, you will learn how to:

- Understand and describe the historical development of technology in the meetings and events industry

- Describe the opportunities and challenges new technologies offer the meetings and events industry

- Understand the key role social media technologies will play in the research, coordination and marketing of meetings and events

- Describe the different types of technological resources available to meeting and event professionals

- Analyze, avoid and resolve the ethical and moral problems that may occur with the incorporation of new technologies

- Describe how the future of meeting and event technology may positively impact your career

- Link the historical development of meeting and event technologies with the future opportunities for improving functionality through the development of new communication platforms, applications and other products

The meetings and events technology field has been painstakingly built by technology luminaries such as Reggie Aggarwall, Corbin Ball, Michael Boult, John Chang, Bruce Freeman, J'Michelle Keller and educators such as Pauline Sheldon, Patti Shock, and many others, whom future generations may also refer to as the crazy ones, the misfits, the rebels and even the trouble-makers. These individuals and their colleagues used their experiences in many other related fields to bring a new industry into being in less than one half of a century.

According to Professor Robert Rausch, former U.S. Secretary of Labor, individuals such as those who worked together bring something new into being may be referred to as symbolic analysts.

"We are living through a transformation that will rearrange the politics and economics of the coming century," says Robert Reich. As we move into the borderless economy, the notion of national products, national technologies, and national corporations will become increasingly meaningless. The only thing that will remain rooted within national borders are the people who make up a nation. This shift has enormous political implications, according to Reich. It means that the traditional idea of national solidarity and purpose can no longer be defined in purely economic terms. It also leads to fragmentation, Reich argues, as "those citizens best positioned to thrive in the world market are tempted to slip the bonds of national allegiance, and by so doing disengage themselves from their less favored fellows." (Rausch,1992)

Traditionally, meetings, events and universities have been referred to as marketplaces of ideas. The marketplace was historically a geographically fixed place. This is no longer true. These marketplaces do not manufacture traditional goods and products in one place; rather they create information, provide education and transform thinking through collaboration in many different places. In order for these global market places to continue to thrive in the future, a new generation of meeting and event technologists is needed who understand the fluent nature of events and will be better prepared to design the applications and software that will promote even greater collaboration in the future.

In little more than half of a century, just about the length of Steve Jobs' life, the meetings and events industry has experienced dramatic change. Table 1.1 provides a brief example of some of these significant changes.

Harry Baum of Great Britain, is a long-time leader in the international meetings, incentives, conventions and exposition industry.

Table 1.1. Paradigmatic Changes in Meetings and Events between 1950 and 2014

FROM	TO
Analog	Digital
Collision	Collaboration
Content	Context
Event	Events without end
Live	Blended
Local	Cloud
Hardware	Software
Human Staff	Technological Stuff

(Goldblatt, 2014)

In the mid 1990s, Harry gave a lecture one evening at the George Washington University in Washington, DC, for a group of sleepy-eyed graduate students. Many of these students had just completed a ten-hour working day and now were in a dark and dusty graduate lecture hall to listen to their guest speaker from England.

Baum began his talk by slowly and carefully announcing in a deep, sotto British voice, "The modern meetings industry was created through warfare." With those profound and intriguing words his audience was soon fully awake and most attentive. Baum explained that most of the computers and aviation innovations used in the modern meetings and events industry that were developed in the previous fifty years were the result of technologies developed for advancing warfare in World War II.

For example, Baum related the development of international airline travel to the need to develop airplanes for long haul troop transportation during World War Two. Further, he stated that electronic computing was developed and subsequently used to fire artillery. According to Baum, the development of the modern global meetings industry was directly related to the development of solutions for creating weapons and strategies for effective warfare.

There may indeed be a technological connection between meetings, events and warfare, and the strange and winding story stretches as far back as the mid-nineteenth century in Scotland. In this chapter we will explore the relatively brief but fascinating traditions and history of technology and will look forward to the future trajectory current and still undeveloped discoveries that may chart the twenty-first century meetings and events industry.

From Analog to Digital: How War Helped Invent Modern Technology for Meetings and Events

It is commonly believed that actual research on finding solutions to equations using mechanical devices started as early as 1836 with a mechanical analog computer called the "differential analyzer." This wheel and disk system may have been one of the first computing devices.

The first description of this device is from James Thomson in 1876. Thomson was from Belfast, Ireland, but lived in Scotland from the age of 10. He called his device the "integrating machine," and with his brother, the famous Lord Kelvin, he published additional descriptions of this early computer. Lord Kelvin later became a renowned mathematical physicist and engineer.

Lord Kelvin recommended that the integrating machine be first used for fire control. Various other scientists were also developing integrating machine technology at this time; however, the first widely practical differential analyzer was constructed by Harold Lock and Vannevar Bush in the United States at the Massachusetts Institute of Technology (MIT) between 1928–1031. The number of integrators used in these machines continued to grow, and scientists in Norway developed a system with 12 integrators in 1938. Only a few years later, once again in the United States, researchers would exceed this functionality through the development of the world's first programmable computer.

ENIAC: The U.S. Army and the Age of Modern Computing

During the 1940s researchers Mauchly and Eckert lead a team at the University of Pennsylvania in Philadelphia to develop a programmable computer that would calculate artillery firing tables for the U.S. Army. In 1946 they announced the development of the first Electronic Numerator Integrator and Computer, or more commonly known as ENIAC.

If you were to enter a room to inspect the ENIAC, you would find huge cabinets containing 17,468 vacuum tubes, 7,200 crystal diodes, 1,500 relays, 70,000 resistors, 10,000 capacitors, and around 5 million hand-soldered joints. The ENIAC unit weighed more than 30 tons and was over 100 feet in length and required a room of 1800 square feet.

The ENIAC computer was programmable, and this was accomplished using punch cards and tape punched with holes. One of the first major tasks required over 1 million punch cards, and the ENIAC was involved in the development of the hydrogen bomb in Los Alamos, New Mexico.

SCREENSHOT

Early Meetings and Events Computer Technology

Although most students today have been using computers since early childhood, the baby boomers (today's students' parents) may only have been exposed to computers in their early thirties. Some graduated from a manual and later electric typewriter to a simple computer manufactured by firms such as Tandy™ which was a brand of the large technology retailer named Radio Shack. Tandy was one of the first companies along with Commodore and Apple to launch the personal computer revolution. In 1976, my first computer was a Tandy 1000. It was a competitor of the IBM personal computer but was more affordable for small and medium enterprise business use. My small events company needed an automated system to maintain a database of suppliers, create and store contracts, and perform simple mathematical computations. The Tandy 1000 solved many of our business problems. However, the on-screen and printed typeface was somewhat fuzzy, it required the use of large stacked sheets of paper with perforated holes on each side (that frequently jammed in the printer), and it took up significant space in our small office. For many, including our small events firm, it would be our first encounter with a personal computer as an essential business tool. Little did we know that this was only the beginning of what would continue today to be a rapidly expanding presence for technology in the meetings and events industry.

Given the early involvement of the U.S. military in the development of programmable computers, it is now easy to understand why Harry Baum remarked that the history of the global meetings and events industry is also the history of modern warfare.

Between the 1940s and the dawn of the twenty-first century computer technology changed dramatically in two dimensions. First, the component parts of computer technology became significantly smaller with the full capacity of the ENIAC system, reduced by 1995 to a silicon chip measuring 7.44 mm by 5.29 mm. This 20 mega-hertz chip was also 20 times faster than the now ancient ENIAC.

Between the 1930s and the final decade of the twentieth century computers became smaller, faster and had much greater capacity for complex computational functions. While the U.S. military may have been the birth mothers and fathers in the development and use of complex computer systems, global business, including meetings and events, became the experienced parents who would set the standards for future productivity through an increased reliance upon technology.

The Real Technology Wars: Hewlett Packard, IBM™ and Apple ™

The Macro Technology Environment

After approaching Bill Hewlett, the co-founder of Hewlett-Packard, for a job in 1968, Steve Jobs incorporated Apple computers only eight years later. Jobs and his partner Steve Wozniak (Woz) developed and marketed their personal computing products Apple I, II and III and Lisa widely. In 1979, Jobs discovered the graphical interface technology developed by Xerox in their laboratories in Palo Alto, California, and adopted this system to dramatically expand the functionality of Apple products.

Just after Apple became a publicly traded corporation in 1981, IBM launched their first personal computer and simultaneously seriously threatened Apple sales. In 1983, *Time Magazine* selected the computer as its machine of the year, much to the disappointment of Steve Jobs who had lobbied and hoped for this honor. During this same period of time, thanks in part to the rapid development of personal computers and the competition between the major manufacturers, the meetings and events industry rapidly adopted computer technology for basic tasks including word processing, registration, finance and accounting and computer-assisted design and drawing to create diagrams and room set-ups.

The invention of the World Wide Web by Sir Tim Berners-Lee in the early 1990s enabled the rapid expansion of communications in the modern meetings and events industry. In 2012 Tim Berners Lee participated in the London 2012 Olympic Games Opening Ceremony and tweeted "This is for everyone," which was simultaneously projected on the backs of 80,000 stadium seats (Friar, 2012).

In 1993 America Online (AOL) and CompuServe combined operations to create a global standard for Internet e-mail service. In the mid-1990s the search engine and social media site that would later become Yahoo was established by two Stanford University Students as "Jerry's Guide to the World Wide Web" (Clark, 2008).

The first weblog was posted by Justin Hall in 1994 and now has resulted in, according to Technorati, 60 million blogs with many targeting the meetings and events industry.

By the turn of the twentieth century, many technology pundits had predicted the worldwide collapse of computer systems due to Y2K (the predicted inability of computers to cope with change in clocks as the end of the millennium). There was, however, little to no disruption reported.

In 2003, Intel Corporation first included Wi-Fi in its Centrino chip, and this greatly accelerated the mobility in computer networking. Wi-Fi would soon be incorporated in the design of most hotels, airports and convention centers.

The phrase Web 2.0 was coined in 2004 by O'Reilly Media to define the rapidly developing interactivity opportunities that were now present within the World Wide Web.

Despite a period of strong economic productivity in the meetings and event sector, a series of dramatic setbacks were looming large on the horizon.

First, in 2008, the financial collapse of the investment banking firm Lehman Brothers created a domino effect throughout the global financial markets, and soon many of the world financial centers including the United States, Great Britain and others were plunged into a widespread economic recession.

In 2009 the huge multinational insurance and financial services firm, AIG, hosted a lavish party while using US government TARP (Troubled Asset Relief Program) funds. The widespread publicity that resulted from this event led to the US government and others chastising the meetings and events industry for irresponsible spending. Even President Obama stated that government meetings should not be held in Las Vegas and, this resulted in an immediate reduction in revenue for this destination. The term "AIG Effect" was coined to describe the new scrutiny meetings and events were receiving by government and other key stakeholders.

However, once again the industry was able to recover by incorporating more technology to compensate for the lack of ability to travel as in times past. Lee and Goldblatt (2012) conducted a study in 2010 of nearly 400 festival event managers, and their findings determined that many event professionals were now turning to social media and websites to replace and lower their advertising costs.

Although the use of holography had first been introduced by Gabor in 1972, it was used for the first time nationally in the presidential election of 2008 as broadcast by CNN. The host, Wolf Blitzer, conducted a life-like television interview with a hologram of the reporter Jessica

Yellen, and it was seen by millions of viewers throughout the world. In 2012 the Coachella Music Festival in California introduced a hologram of the late Tupac Shapur who performed a duet with the very live Snoop Dog before over 100,000 people. According to hologram industry experts, this technology had finally reached its zenith during this astonishingly real performance.

Despite the promise of increasing global technologies, the discovery of high security data being leaked by reporters and whistle blowers Julian Assange, Bradley Manning, and Edward Snowden revealed the further potential for disruption through the expanding accessibility of technology.

The technological developments by the entertainment industry continue to lead the way through film, television, and Internet. One of the latest developments by the Walt Disney Company is titled 3Disney and provides the ability not only to touch but also FEEL in three dimensions as the user touches a textured computer screen. This is accomplished by the process called "tactile rendering of 3D features." According to *The Washington Post* in 2013, engineers at Disney Research developed a rendering algorithm that uses small electronic pulses to trick your fingers into feeling bumps and texture—even though the surface is flat (Dewey, 2013).

Developments in Meetings and Event Technology

During this same period of global advancement in technology, the Braehler company of Germany introduced in 1961 the first single line system for conference communications. Prior to this revolutionary invention dozens of cables needed to be used for conference delegates to communicate with one another. Later, in 1976, Braehler used their rapidly development technologies for 300 delegates to simultaneously communicate at a meeting of the United Nations.

As Apple was introducing the first Mac computer in 1984, Braehler simultaneously introduced European conference delegates to its

DIGIVOTE system. The DIGIVOTE system enabled conference delegates to vote on proposals and candidates from their seating location in the congress center.

A few years later, the trade association Meeting Professionals International (MPI), founded in 1972, introduced just fifteen years later its first Computer Special Interest Group. This group enabled meeting and event professionals to communicate about issues of common concern regarding meeting and event technology. During this same period the first computer-generated name badges made their first appearance and plastic lead retrieval cards were used by trade show exhibitors to track buyer relationships.

In 1988, the hospitality industry recognized they needed a common technology platform, and they developed The Hospitality Industry Switch Company (THISCO). Later THISCO would become Pegasus and result in processing upwards of 300 million transactions each month. The first 1D barcode was also introduced at this time and further advanced the lead generation system for trade shows.

In the late 1980s and early 1990s, meeting and event technology firms such as PC Nametag and Lasers Edge developed software for printing name badges for delegates. In the early 1990s as the first laptop computers began to appear, a new technology entrant, PlanSoft, introduced a web-based tool to standardize meeting and event communications. Due in large part to the rapid development of technology, MPI established the first online discussion group for meeting and event technology professionals, called MPINet.

In the mid-1990s the travel company McGettigan created a web-interfaced software product called Real-Planner. By the end of that decade this product would be further redeveloped and renamed StarCite.

In a unique partnership with MPI and the American Society of Association of Executives (ASAE) PlanSoft began work in 1995 on the first comprehensive searchable facilities database. The final product was finally launched in 1997. During this same period the Holiday Inn Corporation developed and launched the first online system for purchasing hotel rooms. Simultaneously, San Francisco's Miyako Hotel developed the first online request-for-proposal system in collaboration with Cardinal Communications, and in 1996 Motley and Layton founded Pass-Key, the first online group reservations system.

By the end of the 1990s EventSource and StarCite were providing the first online auctions for sale of meeting space. The HotRatesHot-Dates website was the first to offer for sale distressed last-minute hotel inventory to planners in 1999,while seeUthere.com and Eventbrite. com launched the first commercial and consumer meeting and event reservations sites.

By the start of the twenty-first century there were only a few application service providers (ASP); however, they grew in number exponentially during the dawn of this era. As a result of this exponential growth in technology, the first online trade show, ExpoExchange, was conducted in 2000. Delegate networking experienced a huge improvement with the development of ShockFish, which allowed individuals to see pictures and details of other delegates within 20 feet of where they were standing. GetThereDirect launched the first group space reservations tool, and this led to Google and Hotel Planner creating the first small group reservations tools in 2003.

The meetings and events industry was also realizing the need for standardization of processes due to the rapid developments in technology. The Convention Industry Council (CIC) (originally named the Convention Liaison Council) was founded in 1949 as super association (similar to the United Nations) to represent the interests of associations within the meetings and events industry. It subsequently grew to represent over 30 different meeting and events industry associations whose combined membership was over 100,000 people.

In 2000 the U.S.-based Convention Industry Council established a committee to develop accepted practices for the meetings and events industry. This committee developed the Accepted Practices Exchange (APEX), which has been responsible for developing standardized forms and most recently standard reporting practices.

In many ways the turn of the millennium was a major turning point for the development of meeting and event technology. On September 11, 2001, acts of major terrorism occurred in the United States, resulting in the immediate disruption of many businesses, including the meetings and convention industry. Thanks in large part to technological advancements of the recent past, the industry was able to communicate and recover its operations.

Despite the disruptions of 2001, the meeting and event technology industry continued to grow and improve rapidly. E-mmediate was introduced by Hyatt Hotels Corporation in 2002 as an online booking tool for small meetings. The web-based meeting and event networking tools powered by Rio and IntroNetworks were introduced to help people of similar interests connect. Mobile weblogs (McLogs) were introduced during this same time period to rapidly accelerate real-time communications during conferences and exhibitions. At the same time, StarCite and Outtask, Inc. introduced the first real-time airline booking product.

By the early to mid-2000s, new tools were entering the meeting and event technology marketplace. Radio Frequency Identification (RFID) name badges were introduced in 2003 to enable the specific movement of delegates to be tracked by planners.

The Convention Industry Council's (CIC) Accepted Practices Exchange (APEX), the initiative having been first established three years earlier, delivered its first product, the APEX online glossary of terms, edited by Professor Patti Shock of the University of Nevada at Las Vegas.

By 2004, the Hilton Hotels Corporation reported that online bookings for the first time exceeded the reservations received by telephone call centers. This period was also a time of acquisition and merger for the meetings and events technology industry, with PlanSoft and seeUthere merging in 2004 to form OnVantage.

The CIC APEX released its first technology and standardized form product through Toolbox 1.0 in 2005. American Express and Travent Ltd. along with OnVantage cooperated to make tracking of corporate travel much easier and more accurate with the development of the first corporate procurement card.

Two of the oldest meeting and event planning software firms, PeopleWare and Amlink, joined the long list of acquisitions and mergers. The following year, two largest meeting and event technology firms, OnVantage and StarCite, finally merged. At the time of the merger, Corbin Ball, an industry expert, described this as the meetings and events technology equivalent to Microsoft and Apple merging.

The CIC APEX project developed the XML data map in 2006 to move toward the development of an electronic data exchange that would for the first time allow the establishment of true data standards for the meetings and events industry.

The Braehler company once again demonstrated its market leadership through the creation and launch of the first wireless conference system that eliminated the need for even a single line as they established in 1961.

Established meeting and event technology firms such as Cvent, MeetingMatrix, and others continued to grow in mid- to late-2000, with Cvent raising nearly $118 million dollars through its first public offering. MeetingMatrix was sold for an undisclosed sum to hospitality industry technology market leader Newmarket International Inc.

Figure 1.1 depicts the major eras in the development of technology for meeting and events.

70 Years: The Meeting and Event Technological Eras	
1940s	The age of invention: calculation
1950s	The age of progression: computation
1960s	The age of miniaturization: mini computer
1970s	The age of expansion: Apple™
1980s	The age of functionality: meeting and event registration and laser printing
1990s	The age of connectivity: Internet
2000	The age of applications and standardization: APEX XML Data Map
2010	The age of hybridization: MOOCS, Face to Face and Blended Learning

Figure 1.1. Goldblatt Meeting and Event Technological Eras (Goldblatt, 2014)

WHAT IS NEXT?

According to the leading journalists reporting on trends in the meetings and events industry, we are experiencing the tip of a very deep iceberg in terms of future developments in this rapidly evolving industry. Barbara Palmer and Michelle Russell are experienced trade journalists with a combined 50-plus years of experience. Russell is the editor-in-chief of one of the meeting and event industry's most recognized publications, *Convene* magazine, which is published monthly by the Professional Convention Management Association (PCMA). According to Palmer and Russell:

> We need to develop the ability to collect implicit data about our attendees. People read and share differently. Perhaps the further use of RFID badges to create a heat map using Google and, the ability to observe human behavior during meetings and events will

be increasingly important. Also the ability to instantly conduct data through polls and surveys will also be important. The link to education is finding and developing systems to best use your time for what suits your educational needs the best. Perhaps the further development of language translations systems will be dramatically improved allowing for instantaneous communication during meetings and events. (Palmer and Russell, 2014)

Betsy Bair is the experienced content director of a group of meetings and events magazines published under the banner of Meetings-Net. Her magazines include specialist content for medical, insurance, religious and other types of meetings. Bair states that:

> One important innovation would be a surveying application to enable you to instantaneously capture other people's responses to speakers and content. At the 2014 PCMA conference they have invited three students from MIT and three from Harvard to observe the conference and make recommendations regarding the future architecture of this conference. Who knows what brilliant ideas they may come up with that will improve the experience of all participants, both face to face as well as virtual. (Bair, 2014)

These three experienced journalists all agree that while it is difficult to predict the future of technology for meetings and events, the developments will be rapid and will profoundly influence the way planners and their participants engage with content and experience their future meetings and events.

There is a sign over the door of the United States Archives that states "Where Past is Prologue." If the past is our prologue to the future, then one common function of all meeting and event technological innovations appears to be the role of communications. Following the development of the printing press and later the Internet, the most profound development in twenty-first century technology is the mobile telephone.

THE 10-YEAR REVOLUTION OF MOBILE TECHNOLOGY

It may be hard to believe that the first Apple iPhone only became commercially available less than a decade ago in 2007. However, one of its many predecessors, beginning in 1973 was the Motorola model that was the size of a shoebox. Later in 1993, the Bell South IBM Simon model was developed, and this was followed very quickly in 1996 by the Nokia Communicator and in 1997 by the Ericsson GS88, which first coined the term "smartphone." The Ericsson revolution continued in 2000 with the first touch screen and was followed by the development of the Palm mobile phone series (Palm, Palm Treo, Palm One), the first to include a full operational keyboard and to become widely used. Microsoft soon entered the mobile technology race with the Microsoft Pocket, which was the first to use the Windows operating system in mobile technology. In 2002, RIM introduced the Blackberry, which was the first mobile device to allow wireless e-mail capability. In 2007 the iPhone became the first to offer multi-touch interface technology through a highly interactive touch screen. Finally, in 2008, the development and launch of Android system would further accelerate the use of smart mobile telephones throughout the world, with HTC creating the first Android mobile phone.

As a result of these technological developments, mobile phone usage worldwide rose rapidly. For example, between 1990 and 2011 global mobile phone subscriptions increased from 12 million to over 6 billion. Today's modern meeting and event professional relies on the smartphone to communicate, enumerate, and evaluate a wide range of tasks. Many meeting and event professionals consider the smartphone an essential tool for their business use.

The rapid rise of personal computers and mobile technologies has enabled meeting and event professionals to not only work smarter but in some cases also work faster. As computer technology rapidly developed between the 1940s and 1990s, according to McGrattan and Rogerson drawing on U.S. census data, the number of working hours of U.S. workers actually decreased from 41 to 37 hours per week. However, other statisticians contest their analysis and actually state that women were working 300 hours per year more than their male counterparts who were estimated to be working an additional

SCREENSHOT

Early Mobile Communications

A meeting and event company invested in the later 1970s in a device known as a digital pager. This device sent a text message to a box about the size of a pack of cigarettes. The message was usually a telephone number that would need to be called to retrieve a message. Within a few years the digital capacity of pagers grew and more detailed messages could be sent. By the early 1980s we invested in a mobile telephone. The unit was the size of a shoebox, and the one installed in my automobile filled the trunk of the car and required a large exterior antenna affixed to the vehicle. It also required placing the telephone call through a central operator. By the start of the twenty-first century the pager had disappeared and been replaced, thankfully, by the multi-functional and much smaller mobile phone. This allowed meeting and event planners to truly become less reliant on their physical office premises for communications. Throughout each of these technological changes, meeting and event planners have been able to improve their level and quality of customer service as well as increase revenue through more rapid communications. As a result of these developments local, state, and federal laws have been enacted, for good reason, regarding the use of mobile devices while driving motor vehicles. I recall once receiving a mobile telephone call from a large client while driving on a major highway. The client agreed to approve my $500,000 proposal. I was momentarily dazed by this announcement and almost caused an accident. According to the U.S. National Highway Safety Administration, about 6000 deaths and half a million injuries are caused by distracted drivers each year. In 2011, it was estimated that 900 deaths may have been directly caused by mobile phone usage while driving; however, this number could be much higher (National Highways Traffic Safety Administration, 2013).

100 hours. Technology, during the past half century, has either given us an opportunity to work more efficiently and perhaps increase our leisure time or to as shown by the example above concerning using the mobile phone while driving, simply work all the time due to the utility of evolving technologies.

THE EVOLUTION OF THE INTERNET

During the opening ceremonies of the London 2012 Olympic Games, the scientist Tim Berners-Lee was saluted for having helped invent the World Wide Web and then giving it away freely for the world's benefit. The history of the Internet actually began with the introduction of electronic computers in the 1950s.

ARPANET was the first Internet network at the University of California, Los Angeles (UCLA), and the first message was sent from Professor Leonard Kleinrock's laboratory. Later in the 1960s and 1970s, this packet-switched network introduced the development of Internet protocols for linking multiple networks.

The U.S. National Science Foundation (NSF) provided greater access in 1981 to ARPANET through the introduction of the Computer Science Network (CSNET). By 1982 the full Internet protocol suite known as TCP/IP was fully functional, and the introduction of the worldwide Internet was complete.

The 1990s provided revolutionary change in communications technologies through the Internet introduction of e-mail, instant messaging, voice over Internet protocols (VoIP), two-way Internet video, weblogs, social media, online shopping, and of course, the World Wide Web in 1989.

Figure 1.2 depicts Internet penetration worldwide in 2012 and also reveals the digital divide that still exists between the developing and the developed worlds.

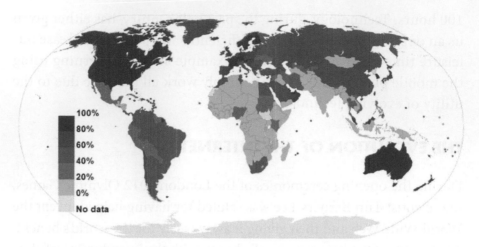

Figure 1.2. Internet Penetration Throughout the World as of 2012 (Source: International Telecommunications Union, 2013, via Wikipedia)

SCREENSHOT

Early Internet Technology

CompuServe was the first major commercial Internet service provider (ISP) in the United States. It was the leading provider of Internet services in a not very crowded field throughout the 1980s. However, a new start-up called America On Line (AOL) disrupted its dominance through the introduction of new fixed rate monthly fees instead of the CompuServe hourly charges. I used CompuServe for my personal home computing at that time. A small electronic modem connected the computer to a telephone line in order to access the Internet. When the modem was connected it was not possible to make a traditional telephone call unless the line was dedicated for the Internet.

One late evening I noticed a string of white text slowly creeping across the black screen. The message stated: "This is Miss Ying in Beijing. How are you?" I stared intently at the screen and then ran to my wife and said, "China is calling!" My excitement reminded me of the first time Alexander Graham Bell placed the first telephone call or Thomas Alva Edison illuminated the first light bulb.

Although the dial-up modem was a slow connection, nonetheless, it was an exciting time where, not unlike the invention of the printing press and later the telegraph, one could expand the footprint of global communications with a few key strokes.

The Internet was in some ways a direct descendant of the digital technology developed in the nineteenth century through the invention of the telegraph. The telegraph was first used as early as 1809 in Bavaria, but the most successful and universally accepted form was demonstrated by the American and New York University Professor Samuel Morse in 1838. Morse demonstrated his ability to send messages using "Morse Code" over a telegraph wire, but it took the U.S. Congress another five years to find a telegraph wire between Washington, DC, and Baltimore, Maryland (about 40 miles distance). Six years later Morse used his device to send the announcement of the political party nominee from midway between Baltimore and Washington, and the message was telegraphed to the U.S. Capitol. In a later message from the U.S. Supreme Court, Morse used his code to ask, "What hath God Wrought?" The technology pioneers from the United States and the United Kingdom who developed Internet protocols with universal standards and later the World Wide Web perhaps did not anticipate the rapid change these innovative technologies would bring to global society.

THE DEVELOPMENT OF SEARCH ENGINES: CANADA, USA AND SWITZERLAND

The first tool used for web searching was called Archie, a name that represents archive without the v. It was invented in 1990 by computer science students at McGill University in Canada. Archie downloaded public file transfer protocol (FTP) data but it did not have the ability to index this data.

In 1991 at the University of Minnesota, Gopher was created, which led to additional search engines called Veronica and Jughead (additional characters in the Archie comic book series). The major advantage of Veronica is that is allowed the use of keyword searching for the first time.

By the summer of 1993, Oscar Nierstrasz at the University of Geneva, Switzerland, created an Internet script called PERL, which led to the development of the world's first primitive search engine in the autumn of the same year. Simultaneously, the world's first web robot was being developed at the Massachusetts Institute of Technology (MIT) in Boston, and this tool was used to generate a web-based index of data that could be retrieved.

In 1994 the first all-text crawler-based search engine called Web-Crawler was developed and was soon followed by the first commercial search engine service, called Lycos. Within the following two to three years search engine names that are familiar today were first introduced. These included Yahoo, which became one of the most popular search engines as users could browse its entire directory as well as use its keyword search function.

Netscape selected five search engines, Yahoo!, Magellan, Lycos, Infoseek, and Excite, to operate in rotation on its web browser commencing in 1996. However, the major entrant was soon to arrive and further disrupt and expand the functionality of search engine technology.

Google arrived in the early days of the twenty-first century, and its major contribution was the incorporation of an algorithm that allowed page rankings (PageRank). Google also featured a simple minimalist search page design that made it very popular with web users.

The primary competitor at that time was a Microsoft search engine product first called MSN Search in the late 1990s and then rebranded in 2009 as BING. MSN Search first used AltaVista as its search engine and then later in 2009, through BING, agreed that Yahoo! Search would be powered by Microsoft BING technology.

Table 1.2. Search Engine Development Timeline (Source: Wikipedia, 2013)

Year	Engine	Current status
1993	W3Catalog	Inactive
	Aliweb	Inactive
	JumpStation	Inactive
1994	WebCrawler	Active, Aggregator
	Go.com	Active, Yahoo Search
	Lycos	Active
1995	AltaVista	Inactive, redirected to Yahoo!
	Daum	Active
	Magellan	Inactive
	Excite	Active
	SAPO	Active
	Yahoo! 2008	Active, Launched as a directory
1996	Dogpile	Active, Aggregator
	Inktomi	Inactive, acquired by Yahoo!
	HotBot	Active (lycos.com)
	Ask Jeeves	Active (rebranded ask.com)
1997	Northern Light	Inactive
	Yandex	Active
1998	Goto	Inactive
	Google	Active
	MSN Search	Active as Bing
	empas	Inactive (merged with NATE)
1999	AlltheWeb	Inactive (URL redirected to Yahoo!)
	GenieKnows	Active, rebranded Yellowee.com
	Naver	Active
	Teoma	Active
	Vivisimo	Inactive
2000	Baidu	Active
	Exalead	Active
	Gigablast	Active
2002	Inktomi	Acquired by Yahoo!

Table 1.2. Continued

Year	Engine	Current status
2003	Info.com	Active
	Scroogle	Inactive
2004	Yahoo! Search	Active, Launched own web search (see Yahoo! Directory, 1995)
	A9.com	Inactive
	Sogou	Active
2005	AOL Search	Active
	Ask.com	Active
	GoodSearch	Active
	SearchMe	Inactive
2006	wikiseek	Inactive
	Quaero	Active
	Ask.com	Active
	Live Search	Active as Bing, Launched as rebranded MSN Search
	ChaCha	Active
	Guruji.com	Active as BeeMP3.com
2007	wikiseek	Inactive
	Sproose	Inactive
	Wikia Search	Inactive
	Blackle.com	Active, Google Search
2008	Powerset	Inactive (redirects to Bing)
	Picollator	Inactive
	Viewzi	Inactive
	Boogami	Inactive
	LeapFish	Inactive
	Forestle	Inactive (redirects to Ecosia)
	DuckDuckGo	Active
2009	Bing	Active, Launched as rebranded Live Search
	Yebol	Inactive
	Mugurdy	Inactive due to a lack of funding
	Goby	Active
	NATE	Active

Table 1.2. Continued

Year	Engine	Current status
2010	Blekko	Active
	Cuil	Inactive
	Yandex	Active, Launched global (English) search
2011	YaCy	Active, P2P web search engine
2012	Volunia	Active
	Cloud Kite	Active, formerly Open Drive cloud search

SCREENSHOT

The Early Search Engines: The Library Card Catalogue

The Bodleian Library at the University of Oxford in the United Kingdom is one of Europe's oldest libraries. A researcher visited the Bodleian to research early graduation ceremonies at the University of Oxford. The octogenarian library research assistant suggested they first search the card catalogue. The card catalogue was a filing system developed in the late-nineteenth century to contain all of the bibliographic records in the modern library. When using this tool, one would manually search thousands of cards filed by the name of the author, title of the record, the subject, and the category.

The research assistant squinted at the handwritten records reaching back to over 150 years of graduation ceremony records. Finally, in exasperation he turned to the visiting as said, "Let's just Google it." The pair adjourned to the research assistant's office and, in a few seconds, found all of the records needed using the Google Drive search engine technology. As the research assistant led the visiting researcher to the actual bibliographic materials needed for his research, he murmured, "These tired ancient eyes really appreciate the Google search engine. The type is brighter, larger and it is easier to access than the card catalogue. Who would have imagined?"

In just a few minutes, the nineteenth century had morphed into the twenty-first century with the Oxford octogenarian research assistant leading the visiting researcher from the past to the future of searching for knowledge.

However, the major advancement in Internet search engine development was still to come. In 2012, through Google Drive searches using cloud-based technology were finally possible. Google described this innovation as "a collective encyclopedia project based on Google Drive public files and on the crowd sharing, crowd sourcing, and crowd-solving principles."

Table 1.2 depicts the historical timeline of the development of search engines.

THE MEETINGS AND EVENTS TECHNOLOGY GOLDEN TRIANGLE

Figure 1.4 depicts the key forces that have contributed to the dramatic increase in technology development and use during the past half century.

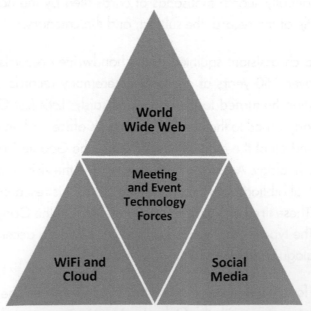

Figure 1.4. Meeting and Event Technology Golden Triangle (Goldblatt, 2014)

ADVANCEMENTS IN SOCIAL MEDIA

The golden triangle of advancement in meeting and event technology must finally include the development of social media. With Facebook now connecting to over one billion "friends" worldwide (a 300% growth in five years) and LinkedIn growing from 0 to over 225 million members in ten years and Twitter tracking over 340 million tweets per day, and finally, YouTube's one billion users watching over six billion hours of video each month, this is a major cornerstone that will promote future grown in this sector.

SUMMARY AND CONCLUSION

During the past half decade, meeting and event technology has rapidly developed and is now moving toward standardization through the influence of APEX and others. Meetings and events are essentially a social as well as educational enterprise and therefore advancements in technology should enhance and not replace the face-to-face experience of delegates. Future innovations will occur in each of the three cornerstones of the golden triangle, as shown in Figure 1.4. These developments will expand the ever-growing technology pie and enable future meeting and event technologists to more quickly reach more people with higher quality experiences at potentially lower costs.

DISCUSSION QUESTIONS

1. How does the history of computer technology impact upon the development of meeting and event management technology?

2. What were three major developments during the period from 1970 to 2000 in meetings and events technology?

3. What are some of the potential future challenges regarding privacy and information protection with regards to meeting and event technology?

TASK

Construct a visual time line using an electronic mind mapping software of the history of meeting and event technology from 1940 to 2014. On the upper level of the timeline list the major milestones in the development of meeting and event technology. On the bottom of the timeline list the major milestones of technology in general as well as key historic world events during this period.

REFERENCES

Bair, B. (2014) Personal communication.

Clark, Andrew. (2008) How Jerry's guide to the world wide web became Yahoo. The Guardian, Feb. 1, 2008.

Dewey, C. (2013) Disney invents touchscreen that lets you feel textures, Washington, DC, The Washington Post

Friar, Karen (28 July 2012). "Sir Tim ls-Lee stars in Olympics opening ceremony". ZDNet. Retrieved 28 July 2012.

Goldblatt, J. (2014) Edinburgh, Scotland, International Centre for the Story for Planned Events, Queen Margaret University

International Telecommunications Union 2013. List of countries by number of Internet users , Wikipedia. Accessed: https://en.wikipedia.org/wiki/List_of_countries_by_number_of_Internet_users. "Percentage of Individuals using the Internet 2000-2012", International Telecommunications Union (Geneva), June 2013, retrieved 22 June 2013

Lee, S. and Goldblatt, J. (2012) "The current and future impacts of the 2007–2009 economic recession on the festival and event industry", International Journal of Event and Festival Management, Vol. 3 Iss: 2, pp. 137–148

National Highways Traffic Safety Administration (2013) NHTSA Survey Finds 660,000 Drivers Using Cell Phones or Manipulating Electronic Devices While Driving At Any Given Daylight Moment, Washington, DC. Accessed: http://www.nhtsa.gov/About+NHTSA/Press+Releases/NHTSA+Survey+Finds+660,000+Drivers+Using+Cell+Phones+or+Manipulating+Electronic+Devices+While+Driving+At+Any+Given+Daylight+Moment

Palmer, B., Russell, M. (2014) Personal communication.

Rausch, R. (1992) The Work of Nations. New York, NY: Alfred A. Knopf World-Wide Web Servers. W3.org. Retrieved 2012-05-14.

Wikipedia 2013. Web search engine. Accessed 8, September 2013: http://en.wikipedia.org/wiki/Web_search_engine)

ADDITIONAL RESOURCES

APEX, Convention Industry Council, accessed: www.conventionindustry.org

"Archive of NCSA what's new in December 1993 page". Web.archive.org. 2001-06-20. Archived from the original on 2001-06-20. Retrieved 2012-05-14.

BITRebels, accessed 7, September 2013: http://www.bitrebels.com/technology/the- evolution-of-smartphones-infographic/

Disney Events Textured Touch Screen, accessed 13, October 2013: http://www.benzinga.com/news/13/10/3985321/disney-invents-textured-touchscreen

Driver Electronic Device Use in 2010." Traffic Safety Facts: National Highway Traffic Safety Administration. December 2011.

GIZMAG, accessed 7, September 2013: http://www.gizmag.com/mobile-pnone-40- year-anniversary-photos/25677/.

Goldstine, Herman H. (1972). The Computer: from Pascal to von Neumann. Princeton, New Jersey: Princeton University Press. ISBN 0-691-02367-0.

Hout, M., Hanley, C., The Overworked American Family: Trends and Non trends in Working Hours, 1968–2011.

"Internet History - Search Engines" (from Search Engine Watch), Universiteit Leiden, Netherlands, September 2001, web: LeidenU-Archie.

Irwin, William (2009-07). "The Differential Analyzer Explained." Retrieved 2010-07-21. "A Century of Difference" Working Paper the Survey Research Center University of California, Berkeley

Isaacson, W. (2013) Steve Jobs. New York, NY: Little Brown.

Oscar Nierstrasz (2 September 1993). "Searchable Catalog of WWW Resources (experimental)".

What's New, February 1994, home.mcom.com. Retrieved 2012-05-14.

CHAPTER 2

Using the Right Technology for the Right Purpose

> *"Think left and think right and think low and think high. Oh, the thinks you can think up if only you try!"*
>
> — Dr. Seuss (Theodore Geisel, 1904–1991)

LEARNING OUTCOMES

As a result of reading this chapter, you will learn how to:

- Identify current and forecast future meeting and event technology needs for your organization through the gap analysis and decision making

- Select the appropriate technological solutions for meeting and event marketing, requests for proposals, planning, registration, on-site management, attendee applications, and evaluation

- Determine the appropriate levels of training that will be required to support these new technologies

- Adapt new technologies for your specific meetings and events purpose

■ Adapt, design, and develop mobile applications for your specific meetings and events purpose

■ Evaluate and analyze meeting and event technological productivity and promote continuous improvement

■ Reduce future meeting and event technology risk through anticipating future technological development

INTRODUCTION

In Chapter 1, seventy-five years of meeting and event technology history were explored to better understand the rapid development of this field of study. This chapter also began to look forward to better understand those emerging technologies that will impact meetings and events in the near future. In Chapter 2, we will examine the various options that you have in selecting the most appropriate technological solutions for your future current and future meetings or events. According to *The Economist* in 2010, Apple announced that their App Store listed 225,000 applications and that they were downloaded five billion times. Therefore, the technological options for potential use in research, designing, planning, coordinating, and evaluating your potential meeting are growing exponentially. As Dr. Seuss, the famed children's author (whose books have sold over 200 million copies to date), suggests, thinking ahead is essential if we are to know which way to turn and not be left behind.

GAP ANALYSIS AND DECISION MAKING FOR MEETING AND EVENT TECHNOLOGY

How do you avoid putting a technological square peg into a meeting or event solution's round hole? The owner of a family hardware store once taught his son that the most important task is first to understand how to select the appropriate tool to perform the job in the most high quality and efficient manner. When the son asked for a practical demonstration of this advice, his father placed before him a hammer and a screw driver and a screw along with a nail. The father asked his son to select the tool to affix the screw to the wall. The son selected the hammer and said, "This will get the job done in record time." The father said "Yes, however, the screw will be damaged and it will not be as securely affixed as if you had taken the time to screw it into the wall." He then demonstrated this action and the negative reaction as the screw bent in half and refused to adhere to the wall.

With the millions of mobile applications and software now available to improve your professional practice and the attendee experience in the meetings and events industry, it is important to know how to select the right tool for the right job. One way this is conducted is through a process developed by Yakov Ben-Haim in the 1980s and known as Gap Analysis Decision Making (Ben-Haim, 2010).

UNCERTAINTY MODEL

According to Ben-Haim, decision making may be seen through three models. The first model is called uncertainty. Starting from the estimate, an uncertainty model measures how distant other values of the parameter are from the estimate: as uncertainty increases, the set of possible values increase—if one is *this* uncertain in the estimate, what other parameters are possible?

MEETING AND EVENT TECHNOLOGY EXAMPLE

If you are planning a city-wide meeting that will require most of the accommodations in a city or region, and these arrangements and contracts must be signed five years in the future, there is a high degree of uncertainty of the technological needs and requirements that may be needed. One example of this future planning was the construction of the Washington, DC Convention Center, when pipes and tubes were planned by the architect for future technologies that has not yet been invented. The architect and convention center planners were indeed looking to the future, one that at that time was uncertain in detailing the types of technologies that would be needed, but certain that they must anticipate future needs.

ROBUSTNESS/OPPORTUNENESS MODEL

The second model of decision making as stated by Ben-Haim is one that assumes there is a certain degree of robustness of information available and opportuneness. Given an uncertainty model and a minimum level of desired outcome, then for each decision, how uncertain can you be and be assured of achieving this minimum level? (This is called the robustness of the decision.) Conversely, given a desired windfall outcome, how uncertain must you be for this desirable outcome to be possible? (This is called the opportuneness of the decision.)

MEETING AND TECHNOLOGY EXAMPLE

If you are planning a recurring meeting or event and have significant data available from which to make a decision, you may wish to make a decision based upon this data. Furthermore, if you have the strong potential of receiving new revenue through commercial sponsorship or in-kind contributions of technology to foster experimentation with technological solutions for your meeting or event, this is known as a desired windfall outcome. One example of this is the wide range of technology providers that support the annual Professional Convention Management Association (PCMA) Convening Leaders conference. Through the data collected over many years by PCMA, they are able to make a robust decision regarding the educational needs of their members and satisfy these needs in part through enhanced technologies. As a result of providing these technological platforms, a large number of technology companies have stepped forward to provide educational programs and demonstrations to enable PCMA members to experiment with new solutions for their business challenges. The windfall of this outcome is increased sponsorship funding for PCMA and new opportunities for their members to explore emerging technologies at low to no cost.

According to Michelle Russell, editor in chief of the PCMA's magazine *Convene,* "We need to collect explicit data from event participants. People now read and share differently, so we need to find new ways through technology to build better profiles about our participants."

DECISION-MAKING MODEL

The third and final decision model is concerned with having sufficient information and the essential time frame in which to make this decision. According to Ben-Haim, in order to decide, one optimizes either the robustness or the opportuneness on the basis of the robustness or opportuneness model. Given a desired minimum outcome, which decision is most robust (can stand the most uncertainty) and still give the desired outcome (the robust-satisfying action)? Alternatively, given a desired windfall outcome, which decision requires the *least* uncertainty for the outcome to be achievable (the opportune-wind falling action)? (Ben-Haim, 2010)

MEETING AND TECHNOLOGY EXAMPLE

In an opportune technology environment, the meeting and event professional would always have the time frame and information needed to make the best decision. For example, when your meeting and event organization is considering refreshing your website, the project scope will specify a period of time to accomplish the tasks required to reach this milestone. Furthermore, within the given time frame the decision will be robust and may result in a desired windfall (such as additional revenues from commercial advertising or sponsorship or philanthropy). Therefore, the decision-making model as argued by Ben-Haim requires sufficient information (research) and an essential time frame (specified to satisfy the aims and objectives of the meeting and event planning organization).

The scientist Stephen Wolfram (2002) further states that decision making is also based upon careful analysis of patterns. Wolfram, who conceived the mathematical software program Mathematica, argues that the mathematical is split into two parts, the kernel and the front end. The kernel interprets expressions (mathematics) and returns the resulting expressions. In "A New Kind of Science," Wolfram explains through his analysis of millions of computations that a problem may be solved and decisions made if it meets the following five characteristics:

1. Its operation can be completely explained by a simple graphical illustration

2. It can be completely explained in a few sentences of human language

3. It can be implemented in a computer language using just a few lines of code

4. The number of its possible variations is small enough so that all of them can be computed

5. Patterns (outliers), processes (community), probable outcomes (change) are predictable. (Wolfram, 2002)

It may therefore be further argued that by applying Wolfram's theory of cellular automata to meeting and event technology, some future technology developments are predictable. Vorob describes this mathematical predictive process as eventology.

Professor Oleg Vorob is known in the field of mathematics as the "'father of eventology." According to Vorob, Goldblatt, and Finkel, eventology is a new direction of probability theory and philosophy that offers the original event approach to the description of variability and ignorance, entering an agent, together with his/her beliefs, directly in the frameworks of scientific research in the form of eventological distribution of his/her own events (Goldblatt et al, 2009).

Further supporting Ben-Haim's decision-making model, the more robust the information, the closer the distribution of factors in terms of decision making. The key application for the meetings and events industry is the importance of robust record keeping by meeting and event planners to reduce uncertainty and to promote robustness and windfall outcomes and lead to improved and more rapid decision making.

Once you have learned how to make the right meeting and technology decision at the right time by using the appropriate tool for your meeting or event task, it is important to link these decisions to actual meeting and event technology products and services. Table 2.1 depicts a few examples of the most common technology decisions that must be made for your meeting or event in each of the five phases identified by Goldblatt (2013).

The number of potential meeting and event technological solutions will continue to grow rapidly. Your task will be to know how to frame the correct questions to ensure that the decisions you make are informed by the best research, evidence, and potential sustainable outcomes for your organization. The following five questions and potential answers in Table 2.2 provide guidance on how to frame these future questions to best serve the current and future technological needs of your meeting and event organization.

Asking the right question at the right time is equally important to finding the right tool for the right job. It is critically important that you take the time to conduct the needed research, to ask the right questions

Table 2.1. Meeting and Event Tasks and Potential Technological Solutions (Source: Goldblatt, 2013)

Meeting and Event Phases and Tasks	Potential Technology Solutions
Research	Online survey systems (SurveyMonkey, MeetingMetrics)
Design	Online room diagramming systems (MeetingMatrix, Room Viewer)
Planning	Online project management systems (Microsoft Project)
Coordination	Online agenda (EventMobi, Quick Mobile)
Evaluation	Online evaluation tools (SurveyMonkey, Zoomerang, and MeetingMatrix)

Table 2.2. Questions to Ask Before Adopting New Meeting and Event Technologies

1. Are the benefits of this meeting and event technology easy to understand and convey to others?

2. Can I point to success stories of other meeting and event organizations who've made a similar decision?

3. Will this meeting and event technology solve a well-known problem in our organization?

4. Will most of our internal stakeholders embrace this new meeting and event technology?

5. Will the meeting and event technology make life easier for me, my colleagues, our suppliers and our attendees?

6. Does your company have the in-house resources to implement the meeting and event technology?

7. Does this technology have a clear return on investment, return on marketing, and return on objective investment (ROI, ROMI, and ROO)?

8. Will our existing systems still work alongside this new meeting and event technology?

9. Will the new technology give our meeting and event organization a clear edge on our competition?

10. Does the meeting and event technology supplier offer training and support for on-ramping and implementation?

(Adapted from Montgomery, 2013)

that will reduce your risk and provide the right model for decision making and ultimately reap the near-, mid-, and long-term benefits that will increase the sustainable outcomes for your meeting and event organizations.

After you have made the decision to adopt the new technology needed for your meeting and event solutions, you must specify

> ## SCREENSHOT
>
> A meeting planner in the 1980s decided to design his own customized event management software to solve several business problems. The problems included maintaining a list of suppliers, managing his clients, and controlling revenue and expense. He hired a computer consultant who devoted many months to creating a customized system. The meeting planner soon learned that there were similar systems were available at a fraction of the cost of the customized program. If the meeting planner had first researched available software and then engaged the consultant to adapt the commercially available software, he would have saved thousands of dollars and been able to use the system several months earlier. Always check and see what software systems may be available before seeking a custom program for your organization.

through each phase of the work flow which tool will best serve your future needs.

The key meeting and event technological tools for event marketing, request for proposals, planning, registration, on-site management, attendee applications and evaluation

According to Silvers (2012) there are six dimensions of the meeting and event attendee experience. These dimensions encompass anticipation, arrival, atmosphere, appetite, activities, and amenities. Technology may positively impact each one of these six critical attendee experiences.

The pre-event research that you conduct using electronic questionnaires may help determine the attendee's future needs, wants, and desires while attending your meeting or event. This research will help you better anticipate their needs. The firm MeetingMatrics describes this phase of research on their dashboard as discovery. Through discovering the preferences of your attendees, you may better plan for the future needs.

Once you have identified your attendees' needs, wants, and desires, you may then proceed to designing the offer that will attract their interest and motivate them to attend your meeting or event. This is the moment when your marketing research conducted through technology is transformed into strategic actions (tactics) through advertising, promotion, and public relations. Each of these tactics is conducted using meeting and event technological systems.

Advertising is the message about your meeting and event that you control. One critical way to control this message is through the design and management of an effective website. The website becomes the central portal through which your prospective and future attendees may discover, explore, recommend to others, and ultimately register for your meeting and event.

Promotion is the opportunity to use an online survey, poll, or other interactive tool to deeply engage with your prospective meeting and event attendees. The results from these surveys, polls, or other tools may then be published in real time on your official meeting and event website.

Public relations is based on research that you disseminate through traditional media, such as radio, television, and the Internet, to further persuade your prospective meeting and event attendees to register. According to Preston (2012), public relations may be used effectively to positively influence your prospective attendees to register for your meeting or event.

While your meeting or event may be marketed to prospective attendees several years in advance, the elements that will produce a successful outcome are acquired through a request for proposal (RFP) process. This process may take place many years in advance if the meeting or event is large in scope or just a few days in advance depending upon the size of your program.

Traditionally, the request for proposal process was conducted by meeting and event planners through a combination of recommendations, research, telephonic interviews, and finally postal solicitations of proposals. Over time, as the Internet rapidly developed, the RFP process moved to a seamless and real-time electronic platform.

Cvent is one online provider of RFP management systems. Their database network of 75,000 meeting and event venues helps create a seamless system for conducting the RFP process. According to Cvent, their planner-friendly supplier search engine provides detailed supplier profiles with extensive facility descriptions, meeting space capacity charts, floor plans, and amenities, as well as exclusive promotions, images of venues, virtual tours and more (Cvent, 2008).

Peppermill Resort in Reno, Nevada, uses MeetingMatrix to allow potential meeting and event planners to download to scale venue diagrams and submit immediate request for proposals. According to Pep-

---

SCREENSHOT

At one time the standard operating procedure for requesting a proposal from a meeting and event supplier was to pick up the telephone and ask for their proposal. The proposal was later sent by facsimile to the meeting planner. This was a cumbersome and time-consuming process. It was also fraught with error because the meeting and event planner's needs could be miscommunicated to the supplier. One time an event planner requested a RFP from a hotel. Within a few days the proposal was received, and following agreement by all parties a contract was issued by the hotel. However, the meeting and event planner used a third party to negotiate the contract; the wrong start time for the opening event was included in the final banquet event order (BEO) at the hotel. The guests arrived early, one hour in advance of the start time on the invitation. Fortunately, the hotel staff sprung into action and received the guests as though they were on time. The electronic RFP process not only promotes efficiency and speed, but it also helps to ensure accuracy through transparency. All parties should have access to the critical information to be able to check and recheck important details such as start times.

permill sales staff, MeetingMatrix allows planners to plan their meetings and events directly online (Peppermill Resort, 2014). The ability to reduce travel expense and time and to receive accurate information in real time is why tens of thousands of meeting and event venue as well as other suppliers have made the transition to the online RFP system.

Dan Berger is the founder and chief executive of the diagramming software solution Social Tables, technology that plays an increasingly important role during the event planning phase. Berger states:

> Today's diagramming solutions (such as Social Tables) are finally liberating the hotel and venue sector from the responsibility of owning the floor plan. These new technological solutions allow planners to create the room set that aligns with their meetings' objectives.
>
> Smart diagramming solutions have a recommendation engine to make intelligent, research-backed meeting design suggestions to the planners using them in order to further drive their ROO (return on objectives).
>
> Taking this concept one step further, future iterations of such software can leverage machine learning to drive meeting effectiveness by comparing survey results to room setups in order to identify the most effective meeting design concepts (Berger, 2014).

Therefore, one key potential outcome of the rapid expansion of technology for event planners is the opportunity to rapidly promote symbiosis between different technological platforms. For example, as Berger suggests, in the near future your participant seating decisions may be directly driven by previous feedback that is recorded in a totally separate but symbiotic technological platform.

What are the appropriate levels of training that will be needed for new technologies?

Every new technological solution should have a prescribed training period to ensure careful universal adoption by all critical stakeholders.

The type of training and its length will be determined by the level of technological sophistication of the stakeholders and the complexity of the new software being adopted. Table 2.3 provides a list of key questions to ask when negotiating training for your new technological programs.

Before scheduling your first training session, conduct an internal assessment of the technology experience and confidence level of those who will be trained. Share this assessment information with your training provider so that the training is offered at the appropriate level. Table 2.4 provides a simple survey instrument to identify the level of staff confidence and skill.

Table 2.3. How to Select the Appropriate Technology Training Provider

1. Know your learning outcomes from the start.

2. Know your software usability from the start.

3. Know your budget from the start; however, build in some flexibility.

4. Know the qualifications and experiences of your trainer.

5. Know how long your trainer has been conducting similar training.

6. Know your options for ongoing customer support.

7. Know the requirements for maintaining your new technology.

8. Know the compatibility of your new technology with other programs and systems.

9. Know about the modularity of your training. Will you only receive the modules you need or must you learn everything all at once?

10. Know the accessibility of upgrades to your current software.

Adapted from Tips for Choosing the ELearning Software That is Right for You (Montgomery, 2013)

Table 2.4. Software Skill Level Survey

(Circle the number on the semantic differential scale that best represents your skill or confidence level.)

---

1. My skill level in the Microsoft Office suite (Word, Power Point, and Excel) is:

Low   1   2   3   4   5   6   7   8   9   10   11   High

2. My skill level in Microsoft Project is:

Low   1   2   3   4   5   6   7   8   9   10   11   High

3. My confidence level in database software systems (data entries and queries) is:

Low   1   2   3   4   5   6   7   8   9   10   11   High

4. My confidence level in running reports is:

Low   1   2   3   4   5   6   7   8   9   10   11   High

5. My greatest concerns about the training I am about to receive include:

_____

When negotiating costs for training there is often a wide range of options. Some firms include some complimentary, free training with the purchase of their software. However, other firms will charge by the participant or a flat rate for the day or entire training period.

SCREENSHOT

"My eyes are killing me!" groaned the meeting and event planner after several hours of training in new event management software. The trainer had failed to enlarge the typeface that was projected on the screen, and after several hours of squinting he had a splitting headache. Looking down at his laptop screen, he noticed that there was a bright glaring gleam that made it difficult to see his work. Between the projection screen and his personal screen, he was slowly going blind. When arranging technology training, try to experience it from the potential students' view and make them as comfortable as possible for the rigorous training they will soon undertake.

Remember, this is always negotiable and you should ask other meeting and event professionals who have used the same supplier about their payment levels and conditions.

Adapting new technologies for specific meetings and events purpose

Several new technologies have been developed in other fields such as Quick Response (QR) codes and Radio Frequency Identification (RFID) chips and scanners. These technologies were first developed to maintain accurate inventory for millions of retail products. In recent years they have been adapted for the meetings and events industry with some degree of success.

It is rarely possible to take a technology from another industry and assume that it will work just as effectively in the highly complex and variable industry of meetings and events. Because the meetings and events industry is essentially the people industry, there will always be nuances and variances as no two delegates are alike.

For example, when using RFID technology some attendees may refuse to wear a chip for religious purposes or simply because they

---

SCREENSHOT

While attending a large convention in Dallas, Texas, the attendee did not realize his name badge included a microchip that was being read by QR scanners throughout the convention center. When he rose to go to the toilet in the middle of a large plenary session, his name suddenly flashed on the large screens at the front of the room. His name badge had signalled the scanner that he was leaving the room and he was embarrassed as several attendees laughed at his grand exit. When using QR codes and scanners, notify attendees during the registration and ask them to tick a box confirming their acceptance of the microchipped badge. Always let them know what you will be doing with the data you collect and how it will be protected in the future.

do not wish to be followed and tracked by the ubiquitous scanners throughout the venue. Still other attendees may not wish to use their mobile device to scan a QR code as this requires an additional step in their delegate experience. Therefore, before introducing these technologies into your meeting or event, it is important to conduct pilot tests and receive feedback from your attendees so you may make any necessary adjustments prior to the full roll out of the technology for all attendees.

Table 2.5 provides examples of how to use QR codes and RFID scanners effectively to manage your meeting or event.

There are now hundreds of thousands of mobile applications now available, and to many meeting and event planners the choice of an appropriate application for their meeting or event may appear to be confusing. However, this process may be simplified by first following these guidelines.

Table 2.5. Opportunities for Using QR Codes and RFID Systems

1.	The QR code may be used to provide detailed information from a website for the delegate about poster presentations, new products and services at an exhibition, and other critical elements within your meeting and event.
2.	The QR code may be used to provide the attendee with additional sources, references, and other material that will enrich their learning experience.
3.	The RFID system may be used to monitor attendance at individual sessions.
4.	The RFID system may be used to monitor the flow of your attendees through the meeting and event.
5.	The RFID system may be used to monitor the interest levels of your attendees by tracking which trade show booths they visit and for how long they remain in each booth.

(Adapted from Montgomery, 2013)

Always begin by surveying your attendees to determine the types of technology they are using (Android or iPhone).

Ask your attendees about their needs, wants, and desires for technology before, during, and following the meeting or event.

Finally, investigate the Wi-Fi capacity for your meeting or event venue in advance to determine if your attendees can easily to access your applications.

SCREENSHOT

A medical conference was held in Barcelona for over 10,000 doctors, scientists, and researchers. The professional conference organizer (PCO) decided to use a mobile application for the attendees to access the program. This would be the first ever paperless program for this conference.

During the first hour of the conference 10,000 attendees received ERROR messages on their mobile application. Following an investigation, the planners discovered that the number of users had exceeded the Wi-Fi capacity for the venue.

To resolve this problem, the planners immediately sent an e-mail to all users with a PDF attachment of the printed program. Although the attachment was not searchable, it did allow the attendees to find their seminar rooms.

For application access your attendees will need 3G or 4G connections throughout their mobile devices (telephones, tablets, laptop computers). According to Diffen.com the definition of second generation, third generation, and fourth generation, commonly known as 2G, 3G and 4G, encompasses the following historical development: 3G and 4G are standards for mobile communication. Standards specify how the

Table 2.6. Development and Capacity of 3G versus 4G Technology

	3G	4G
Data Throughput	Up to 3.1 Mbps with an average speed range between 0.5 to 1.5 Mbps	Practically speaking, 2 to 12 Mbps (Telstra in Australia claims up to 40 Mbps) but potential estimated at a range of 100 to 300 Mbps.
Peak Upload Rate	5 Mbps	500 Mbps
Peak Download Rate	100 Mbps	1 Gaps
Switching Technique	Packet switching	Packet switching, message switching
Network Architecture	Wide Area Cell Based	Integration of wireless LAN and Wide area.
Services and Applications	CDMA 2000, UMTS, EDGE etc.	Wimax2 and LTE-Advance
Forward error correction (FEC)	3G uses Turbo codes for error correction.	Concatenated codes are used for error corrections in 4G.
Frequency Band	1.8–2.5 GHz	2–8 GHz

airwaves must be used for transmitting information (voice and data). 3G (or 3rd generation) was launched in Japan in 2001. As recently as mid-2010, the networks for most wireless carriers in the U.S. were 3G. 3G networks were a significant improvement over 2G networks, offering higher speeds for data transfer. The improvement that 4G offers over 3G is often less pronounced. Analysts use the analogy of standard vs high-definition television to describe the difference between 3G and 4G (Diffen.com, 2013). Table 2.6 compares the development and capacity of 3G versus 4G technology.

Wi-Fi

According to The Wi-Fi Alliance, Wi-Fi is defined as any wireless local area network (WLAN) that are based upon the Institute of Electrical and Electronics Engineers 802.11 standards. Wi-Fi is therefore a synonym for WLAN (Webopedia, 2014).

According to Scott Reeves of Ruckus Wireless, to deliver seamless Wi-Fi for 10,000 persons, you must consider using Wi-Fi applications positioned throughout the venue.

> If you're expecting 10,000 or even 20,000 people to attend a huge exhibition, you can also be fairly sure that not all of them will be accessing the wireless network at the same time. If you think about the WOMAD (World of Music, Arts and Dance) festival that took place at the end of July, the event attracted some 40,000 visitors, but there were never more than 500 concurrent users on the Wi-Fi network, with around 2,000 unique users in total. Traffic on the guest network never exceeded the 100MB backhaul connection.

> At the WOMAD festival, we were able to guarantee reliable Wi-Fi access for both guests and organizers using 35 APs (access points) positioned at various points throughout the 350-acre site, with more near the main stage and in other densely populated areas. This meant that each AP had to support no more than 100 users at any given time; with the back-up, it could scale to support 200 users during peak times. Load balancing like this is key to ensuring that no single AP becomes overloaded and impacts on the user experience. Another factor to consider is RF (radio frequency) management—using beamforming technology, it is simple to maximize throughput from each AP and minimize the impact of interference from neighboring devices. (Diffen, 2013)

Therefore, when planning to incorporate mobile applications within the context of your meeting or event, extensive pre-planning is required to ensure seamless delivery. You may also wish to incorporate signage as shown in Figure 2.1 below to identify Wi-Fi hot spots where connectivity is stronger for your attendees.

Figure 2.1. Wi-Fi Hot Spot Universal Icon Sign

Once you have established that your guests are able to connect easily with their mobile applications, it is important to consider the type and range of applications that you may wish to use for your meeting or event.

The first question you must ask is why you want to incorporate an AP or several APs into your meeting or event experience. For example, do you wish to try and reduce your event's carbon footprint? Do you want to use an AP to increase the convenience level for your attendees (an electronic program versus a larger bulky paper one)? Do you want to send (push) announcements to your attendees? Finally, do you want to increase the networking capacity among your attendees before, during, and after the meeting or event?

According to meeting and event technology expert Corbin Ball (2014), Certified Meeting Professional (CMP) and Certified Special Events Professional (CSP), you should incorporate the following considerations into your decision making:

- Data connection: All apps should include an offline mode, meaning no data connection is needed once the application is installed or saved within the browser. However, for updates or push notifications, the apps require some data connection. If a connection is not available onsite, a hybrid or native app would be more stable on most phones.

- User experience: Because hybrid and native apps are downloaded, many prefer their faster, smoother access to information such as speaker bios and agendas. For polling, audience response and survey tools, they also provide a better user experience. If there is no signal when taking surveys, for instance, both hybrid and native apps hold the data and send it when there is a connection.

- ■ Community involvement: Developers of hybrid and native apps have made great strides in social interaction by allowing push technology to be used within the apps. For example, many developers now embed the use of the device's camera within the app to post pictures more efficiently than web apps.

- ■ Cost: While hybrid and web-based apps are typically faster to deploy and less expensive, the newest technology for these options have made it possible to develop apps that offer many of the same experiences as native apps. So if you do not require advanced programming, it is likely the best way to go. Native apps tend to cost more because they require custom coding of features, use of the device hardware, and GPS (Global Positioning System).

- ■ Design elements: The beauty of apps in today's market is that look and feel, branding, menu/navigation, and sponsorship options can all be customized and can function similarly no matter the app platforms.

- ■ Content: Both static content and dynamic content, such as registration, can be formatted for hybrid, web-based, and native apps. Attendees' needs and planner recommendations should determine the app content. At a minimum, most audiences expect agenda/speaker information and meeting room maps.

- ■ Audience demographics: Knowing your audience is crucial when deciding on what app to deploy. If you are in an industry that relies heavily on tablets or iPads, your audience is probably most familiar with native apps. If this is the case, hybrid apps can be a great option to offer a native app experience without the custom programming cost. In contrast, if your audience is a late adopter to mobile devices, using a browser to view the content is likely preferred. (Ball, 2014)

Once you have considered these key points, then you should begin to consult with a wide variety of application providers to identify those that provide the best value for your investment and offer the strongest technical support for your attendees. It is important to note that every day hundreds or perhaps thousands of new applications are being developed. Therefore, you should proceed carefully to identify the best provider for your meeting and event today as well as for the next few years. Attendees prefer to use an application that is familiar rather than having to learn a new application every year.

The main aim in adopting new technology for your meeting or event is to increase overall productivity for planners and their attendees. This may be accomplished through continuous evaluation and analysis to demonstrate improvement year upon year.

Continually evaluating and analyzing meeting and event technological productivity to promote continuous improvement

There are several methods for evaluating and analyzing meeting and event technological productivity. Traditionally, this measurement has been decentralized and therefore the results are not always valid and reliable. Figure 2.2 depicts a model for comprehensively evaluating

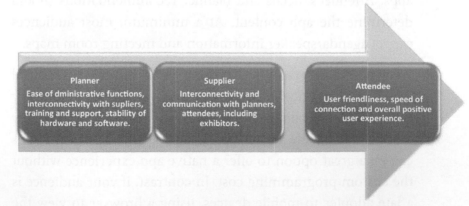

Figure 2.2. Comprehensive Evaluation Elements of Various Stakeholders (Source: Goldblatt, 2014).

meeting and event technological productivity that will join up all of the feedback from the stakeholders within your organization.

The methods that you use to evaluate these outcomes may vary widely depending upon the individual needs of your meeting or event. However, it is important to avoid the mistake of looking for simple solutions to this process. The potential hierarchy of meeting and event technological evaluation is depicted in Figure 2.3.

As Figure 2.3 depicts, the user engagement outcomes are the highest level of evaluation desired from the measurement process. How many times do your users engage with the technology you have introduced and how long to they engage each time they log in? The next level of measurement is the user experience. How easy and friendly do your users find the new technology you have introduced? Finally, while connectivity and speed are important metrics of evaluation, this is only the gateway to the higher levels of experience—which is experience and engagement.

The introduction of new meeting and event technologies brings new risks for your meeting and event organization. However, nothing

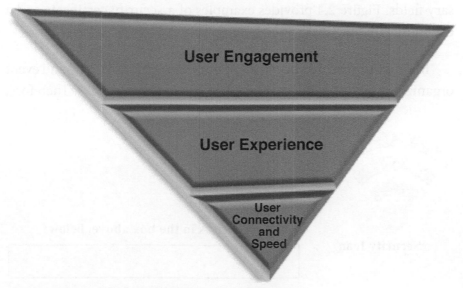

Figure 2.3. User Technology Hierarchy of Evaluation

ventured, nothing gained. Indeed, a new technological advancement is a new venture for your organization and should be welcomed as an opportunity to find new and better ways of delivering a better experience for your staff, suppliers, and attendees.

Traditionally risk is managed by controlling three variables. The term ART means avoid, reduce, and transfer and may be applied to your technology decisions.

You may wish to avoid certain risks by not using certain technologies. For example, if you are using Wi-Fi technology in a public building, you may wish to avoid using unsecure public servers and instead use a secure, private server to help ensure your data is not available to others.

However, you may be able to reduce some risks by incorporating your meeting and event technology supplier as a consultant through the testing process. Furthermore, you may wish to include a security statement on your registration page notifying your attendees that their information is being encrypted for security purposes. Finally, you may wish to use a decoding device to authenticate that the person signing into your registration website has the capacity to complete the necessary fields. Figure 2.4 provides examples of a security certificate and a decoding window.

Internet security has become a major concern for meeting and event organizations due to the increased number of cyber attacks. Therefore,

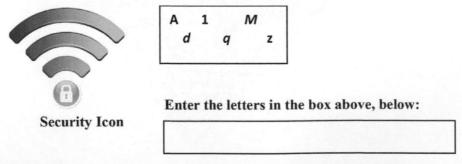

Figure 2.4. Examples of a Security Certificate and Decoding Window

reducing the ability of unscrupulous individuals to access your website and information infrastructure is extremely important.

The third and final method of controlling risk is to transfer some or all of the risk to others. You may do this with a statement on your website that limits your risk and transfers some or all of the risk to the user or to your security partner. You may wish to include a statement with a tick box as show below that informs the user of their future risks, such as:

Every reasonable effort has been made to protect your security. However, it is impossible to guarantee one hundred percent security. Therefore, the user understands and accepts that by submitting their data they are also accepting these inherent risks.

❑ **Tick the box above to confirm your understanding and acceptance of this risk.**

It is important to note that many countries and regional governmental bodies, most especially the United Kingdom and the European Union, have strong data protection laws that may require more stringent protections be provided by meeting and event planners when accepting and guarding attendee data. (For more information, visit: http://ec.europa.eu/justice/data-protection/index_en.htm.)

SCREENSHOT

In 2000, as a result of the clocks changing due to the start of the new millennium, many insurance companies specifically excluded computer technology as a risk they would cover in the instance of a technological malfunction at midnight.

Even today not every insurance policy will cover data security and other issues related to meeting and event technology. Therefore, before you transfer the risk to an insurance company, check carefully for all exclusions that may involve your technological assets.

SUMMARY AND CONCLUSION

Selecting and effectively using the appropriate technology for your meeting and event is one of the great challenges of the twenty-first century for planners. However, the range, scope, breadth, and depth of applications and solutions that are now available helps ensure that you will find the right tool for the right job. Once you have selected the right tool, it is essential that you learn how to use it confidently, and that is why training and support from your technology provider is extremely important. Finally, with new emerging technologies come new risks. Therefore, you must avoid, reduce, and transfer these risks whenever possible to reduce injuring your attendees as well as diminishing the reputational advantage of your organization.

DISCUSSION QUESTIONS

1. How do you know which technology tools will provide the most efficient and effective outcomes for your meeting and event organization?

2. How do you select a proper technology provider and assess their ability to provide training and support?

3. How do you avoid, reduce, and transfer the risk when adopting new meeting and event technologies?

TASK

Create a request for proposal (RFP) to submit to three technology providers (suppliers) to provide a meeting and event registration application for use by your attendees. Your attendees are located in ten different countries and use three different currencies (British Stirling, Canadian dollars, and U.S. dollars).

REFERENCES

Ball, C. (2014) Tech Talk Newsletter. Accessed: http://www.corbinball.com/techtalk/

Berger, Dan (2014) Personal Communication.

Ben-Haim, Y. (2010) *Info-Gap Economics: An Operational Introduction,* New York, NY: Palgrave-Macmillan

Cvent (2008) *Cvent Launches Event Venue Directory and RFP Management System,* accessed 17, February 2014, viewed: http://www.Cvent.com/en/company/rfp-management-system.shtml

Diffen.com (2013) *3G vs 4G.* Viewed 24, February 2014. Accessed: http://www.diffen.com/difference/3G_vs_4G

The Economist, (2010) *Apps and downs,* accessed 16, February 2014, viewed: http://www.economist.com/node/16381330

Goldblatt, J. (2013) *Special Events, Creating and Sustaining a New World for Celebrations,* New York, NY: John Wiley & Sons, Inc.

Goldblatt, J. (2014) International Centre for the Study of Planned Events. Edinburgh, Scotland, Queen Margaret University.

Goldblatt, J., Finkel, R., Vorob, O, "Eventological Theory of Decision-Making", J. Sib. Fed. Univ. Math. Phys., **2**:1 (2009), 3–16

Montgomery, I. (2013) *Ten Questions to Ask Before Adopting New Technology, VROOZI,* Accessed 16, February 2014, viewed: http://vroozi.com/blog/10-questions-ask-adopting-new-technology

Peppermill Resort (2014), *MeetingMatrix: Creating a Seamless and Stress Free Environment for Your Guests,* accessed 17, February 2014, viewed: http://www.peppermillreno.com/meetings-weddings/meeting-planner-tools/meetingmatrix

Preston, C. (2012) *Event Marketing, Second Edition,* New York, NY: John Wiley & Sons, Inc.

Reeves, S. (2010) *Delivering Wi-Fi for 10,000 Delegates.* Mobile Industry Review. Viewed 23, February 2014. Accessed: http://www.mobileindustryreview.com/2010/08/scott-reeves-of-ruckus-wireless-on-delivering-Wi-Fi-for-10000-delegates.html

Russell, M. (2014) Personal communication.

Silvers, J. (2012) *Professional Event Coordination,* New York, NY: John Wiley & Sons, Inc.

Webopedia (2014) Viewed: 24, February 014. Accessed: http://www.webopedia.com/TERM/W/Wi_Fi.html

Wolfram, S. (2002) *A New Kind of Science,* accessed 16, February 2014, viewed: http://www.wolfram.com/products/

CHAPTER 3

Meeting and Event Search Engine Optimization

> *"Social is the way our work gets discovered. Content that is truly exceptional, unique, and useful can earn tremendous awareness through social media, and that social amplification often leads to great links, which leads to great rankings."*
>
> *Rand Fishkin, CEO/Co-Founder, Moz (2014a)*

LEARNING OUTCOMES

As a result of reading this chapter, you will learn how to:

- Think and why to think about SEO before launching or redesigning a website for an event

- Create a content strategy, which is one of the top three elements in a SEO campaign

- Create an inbound links strategy—the second most important element in a successful SEO campaign

- Use social media strategy in favor of every SEO campaign

- Choose and use keywords and phrases in SEO

■ Integrate SEO thinking in all communication efforts

■ Building a strong event brand will help your SEO strategy

INTRODUCTION

Let's start with one question: Where do you go if you need information? Without any doubt the majority of people today will open a search engine site. Billions of searches are conducted every single day on popular search engines and social networking websites by people all around the world. For many that is not only the first place for information, but the only one, so we have to do our best to make our sites visible and accessible if we want our potential guests to find us quickly and easily. "Google continues to dominate the list of most used search engines. Asked which search engine they use most often, 83% of search users say Google" (Purcell, Brenner, and Rainie, 2012). However, this is not the only important thing to bear in mind. Studies show that most users, in fact, skip to the organic results, without paying attention to the advertisements on the top. Usually the first three results receive attention. This phenomenon is called the *Golden Triangle*. Of course, it is difficult to get there, but not impossible.

Around 83% of the people are a serious enough reason to start thinking about SEO long before we start the communication of our event. This is especially true if the event website plays a pivotal role in selling tickets, getting new registrations, or spreading the word for your event. SEO is an acronym for "search engine optimization". It is the process of increasing the number of visitors to a website by achieving high rank in the search results of a search engine. The higher a website ranks in the results of a search, the greater the chance that users will visit the site. It is common practice for Internet users not to click past the first few pages of search results, therefore high rank in SERPs (search engine results pages) is essential for obtaining traffic for a site. SEO helps to ensure that a site is accessible to a search engine and improves the chances that the site will be indexed and ranked favorably by the search engine (LaFerney, 2007). It is obvious that nowadays we cannot ignore the importance of having high rank. Although there are numerous tactics to achieve this, most specialists agree that interesting and involving content is still a very important element for success.

THE ANSWERING MACHINE

From users' point of view search engines have to provide answers—quickly and precisely. Most of us have neither the time nor realize the necessity to reflect on how to approach this task. What we want are answers. We have to remember that the primary responsibility of those engines is to offer relevant results to their users. If they manage to do it, users will come back again and again when they need answers. To be able to supply us with answers, the right ones, they have a very complicated system, which will not be discussed here. However, we will presently talk about the steps we need to take, considering some small details in order to receive a better rank.

If you consider more carefully what you are searching for, you will come up with the following three main reasons:

■ To do something: You want to buy tickets, clothes, books, movies etc. You know what you want, but you need either a better offer or a cheap or fast delivery.

■ To gain further knowledge: You need information about a topic, problem, etc.

■ To open a particular page: With all those new technologies no one can remember the address of the pages they visit.

That is why it will be not exaggerated to say that many of us do those things every single day. Now we can search even for pictures, to find out who is in the picture, get the name of a song. In fact, search engines influence not only our online behavior but what we do offline as well. So, if we want more guests at our event, we have to do our best to come out on top in search results.

WHAT SEO CAN DO

With the development of technology the algorithm of these machines changes. Yet, there is something that comprises a significant part and influences results—the popularity of the page. If a lot of people spend time on your page, if they come back and share with their network, that means popularity (or importance), and search engines want to offer that great page to its users. Search engines offer information about their rules. See Figures 3.1 and 3.2 for search engine sources and their limitations.

Google: https://support.google.com/webmasters/answer/35291?hl=bg

Bing: http://www.bing.com/webmaster/help/webmaster-guide-lines-30fba23a

Yahoo: http://www.seo-yahoo.info/

Figure 3.1. Additional Information about SEO

#1 If you use an online form, accessible after login, it cannot be reached by search engines.

#2 If you duplicate your content on your site, blog, Facebook page, etc., the search engines will ignore it, as they are searching for unique content.

#3 If you use Flash files, photos, videos, audios, you have to know that they are still difficult to be indexed by search engines.

#4 In case you decide to be original and you do not use common terms in your content, search engines will not be able to understand you. They are still just engines, not humans.

#5 If there are no links to your pages, the search engines probably will ignore you. That means you are not popular/important.

#6 If you put a lot of links on one page, that means you will put this page at risk not to be indexed.

Figure 3.2. A Few Important Limitations for Event Planners

Of course there are limitations. In Fig. 3.2 you can see some of the most important limitations, according the event management profession.

It is important to know exactly what to expect from SEO, what we can do better, and what has to be done by experts. The difficulty here is that each single page of your website should be easy to use both for consumers and search engine robots. In many cases not being specialists, we tend to focus more on the user's experience than on the search engine itself. The main goal of SEO is to make each page accessible and readable for robots. Since robots collect information about your pages, if the information is not given in the proper way, it will be ignored.

THE BASICS OF SEO

The father of modern public relations, Edward L. Bernays, said that in times of mass communication, to be modest is a personal virtue and a public vice. Therefore, we have to do our best to let people know about our event. Some experts call search engine specialists the lobbyists of the future because assisted by them you can be on the top of organic results—those that are not paid. They know much better than we do how those engines work.

We can find out how search engines see our site by using seo-browser.com. The simple analysis is free. It is enough to get an impression of what you have done right, or what is wrong. Sometimes to have a great look means to be invisible for the search engines. The best combination is to have plain text, with your keywords, and pictures or video that are supplementary to the content. Do not rely only on visuals.

WHAT YOU CAN DO FOR BETTER SEO

Of course, the best way is to have a budget for working with the best professionals in order to get the best results, though sometimes that is not possible. We hope that you will have the opportunity to work with the top specialists. However, if in some cases you have to count on yourself, here are some helpful insights.

KEYWORDS

"It's only words, and words are all I have to take your heart away". You may remember those words from the Bee Gees song. We can say that regarding all technologies and advances in computer programming: words, especially keywords, are still the main tool by which people search. That means we have to walk in users' shoes and see how to use those keywords in the best possible way. Since good, attention-grabbing content is the most efficient way to get a good ranking, we have to focus on that, as we are responsible for writing the content for our website. It goes without saying that keywords are fundamental for searching. The first thing we do when we want to search is type a combination of words in the browser. So, if you want your page to appear in a higher position for "events", you have to be sure that the word "events" is part of your content that will be indexed. The first thing you have to do is to think about your keywords. After that, it is a good idea to try to imagine what words your potential guests will be using to search about

> **DEFINITION:** Metadata describes other data. It provides information about a certain item's content. For example, an image may include metadata that describes how large the picture is, the color depth, the image resolution, when the image was created, and other data. A text document's metadata may contain information about how long the document is, who the author is, when the document was written, and a short summary of the document.
>
> Web pages often include metadata in the form of meta tags. Description and keyword meta tags are commonly used to describe the web page's content. Most search engines use this data when adding pages to their search index. (Techterms.com, 2014)

your event. It is precisely from those two lists that you have to make up your own keywords. If you decide to choose general keywords as event, special event, conference, your work will be very difficult, as everyone in the industry is using them. Moreover, there is a very small chance for users to search for those general terms without any additional word. So the best idea is to choose more specific words that describe your unique event and use them naturally in the text on the pages of your website. The best you can do for the optimization of your pages is to make sure keywords are used naturally in titles, texts, and metadata.

Relying on our experience as SEO specialists, we would suggest that you take the following steps:

■ **Put the keyword in the title tag**

A title tag, the main text that describes an online document, is the second most important on–page SEO element (the most important being overall content). It appears in three key places: browsers, search engine result pages, and external websites (Moz, 2014b). Careful choice and application of keywords is crucial because the closer the result to the first page is, the greater chance you stand to be spotted by users.

Usually search engines have a limited number of characters that appear in the results. At the moment this number is between 65 and 75. The rest will appear as ... So if you want users to have a chance to see your keywords, you have to fit your title tag in these limitations. Except for the precisely chosen key-words, if your brand has good reputation, you have to use it here, as people are more likely to trust brands with which they are already familiar. In terms of event management, the title tag can be viewed as an invitation, the first time your potential guest has any contact with your event. We are well aware that there is only one opportunity to make a first impression, so we cannot afford to spoil it. We have to put an effort to make this first meeting as emotional as possible, because even though people are rational, in most cases they tend to take decisions emotionally.

■ **Use it in the first paragraph**

Text is, in fact, the content people are looking for. That is why including the information users most probably need is a must. A frequent mistake we are inclined to make is to write the text only to meet the requirements of our boss. Yet it will not be the boss who will be inquiring about the event in different search engines, so we have to write for those who are unfamiliar with our event and who will be willing to join in after reading the text. Consider carefully what kind of information your potential guests will need in order to come to the event. Give them at least three reasons to attend. Thus, writing your compelling text, you have to use your keywords as early as possible.

■ **Say it 2 or 3 times in the body of the text**

The power of words gives us the opportunity to use them in different combinations and variations. Bearing this in mind, we have to integrate words naturally into our content wherever they fit best. Talking about keywords, we can include phrases too. While working on our first presentation of the book, we decided to use "Events without end" as keywords, which we then repeated in various parts throughout the book. Each of us had the freedom to contribute by making our own suggestions. In the end, all guests were well aware that today's events are without end (just one of all possible variations of our keywords).

■ **Include it in the URL address**

No doubt everyone has seen URL like this: http://tedxnbu.com /%D1%80%D0%B5%D0%B3%D0%B8%D1%81%D1%82%D 1%80%D0%B0%D1%86%D0%B8%D1%8F-tedxnbu-2014/.

According to all criteria it does not look good at all, so the creation of such a URL must be avoided. Although most web interfaces give us opportunities to write our URL, we do not frequently avail ourselves of these possibilities. In case you find it difficult to write your precise URL every time, think of using software for short links, which has at least two benefits: (1) the link looks much better (for example, http://goo.gl/JMvidk), and (2) you have access to see where people clicking on your link come from. See Figure 3.3 for sites that can provider shorter URLs.

Fortunately, there is technology to help us in every single task we want to do. Moreover, there are sites which can help you with writing good-looking, relevant URLs. You can use Apache Module mod_rewrite, which provides a rule-based rewriting engine to rewrite requested URLs on the fly (http://httpd.apache.org/docs/current/mod/mod_rewrite.html) (see Figure 3.4) or ISAPI_Rewrite, which is a powerful URL manipulation engine based on regular expressions (see Figure. 3.4). Although its functions are quite similar to those of the Apache's mod_Rewrite, it is specifically designed for Microsoft's Internet Information Server (IIS).

Name of the site	URL	Opportunity to track clicks
Google URL shortener	http://goo.gl/	Yes
Bitly. The power of the link.	https://bitly.com/	Yes
Ow.ly - Shorten URLs, share files and track visits	http://ow.ly/url/shorten-url	Yes
Tiny URL | a Simple URL shortener	http://tiny.cc/	Yes
Bit.do URL shortener - Shorten, customize, and track your links	http://bit.do/	Yes

Figure 3.3. Creating Short Links

The mod_rewrite module uses a rule-based rewriting engine, based on a PCRE regular-expression parser, to rewrite requested URLs on the fly. By default, mod-rewrite maps a URL to a file system path. However, it can also be used to redirect one URL to another one, or to invoke an internal proxy fetch.

Mod_rewrite provides a flexible and powerful way to manipulate URLs using an unlimited number of rules. Each rule can have an unlimited number of attached rule conditions, to allow you to rewrite URL based on-server variables, environment variables, HTTP headers, or time stamps.

Mod_rewrite operates on the full URL path, including the path-info section. A rewrite rule can be invoked in httpd.conf or in .htaccess. The path generated by a rewrite rule can include a query string, or can lead to internal sub-processing, external request redirection, or internal proxy throughput. (Apache Software Foundation, 2014)

Figure 3.4. Description of Mod_Rewrite

What you can do with ISAPI-Rewrite:

- Optimize your dynamic content, like forums or e-stores, to be indexed by popular search engines.

- Block hot linking of your data files by other sites.

- Develop a custom authorization scheme and manage access to the static files using custom scripts and database.

- Proxy content of one site into a directory on another site.

- Make your Intranet servers accessible on the Internet by using only one Internet server with very flexible permissions and security options.

- Create dynamic host-header based sites using a single physical site.

- Create a virtual directory structure of the site hiding physical files and extensions. The latter also facilitates switching from one technology to another.

- Return a browser-dependent content even for static files.

What's more, a wide range of other problems could be solved if we take advantage of the regular expression engine built into the ISAPI-Rewrite. (Helicon Tech, 2015)

Figure 3.5. Description of ISAPI–Rewrite

■ **Use it in meta tags**

In general, meta tags have to provide search engines with information about the content on the website. It is obvious from Figure 3.6 that all we can see while searching are meta-tags: meta titles and meta descriptions. That's the reason to give them all due attention. In most cases all a search engine will show is the first 155 characters of your meta description, so try to fit all you want to say in that space. Provided that you are used to the Twitter standard of 140, 155 will be quite sufficient.

In writing a meta title tag your goal should be to fit within 65 characters. Everything after this number will appear as … You can always search for help online. As for meta title tags and meta descriptions, you can use SEO tag counting tools. Just do a search on those keywords to see how many of the available suggestions you will find convenient to use.

"Words are all I have to take your heart away"

Now let's talk a little bit more about words. In May of 1929, General Electric and Westinghouse approached Edward L. Bernays with the

Figure 3.6. How SEO Works in Google Search

task of handling the 50th anniversary of the first incandescent light, a celebration which would honor both Thomas Edison and his invaluable invention. Edward L. Bernays spent a lot of time searching for the right words to name the event. Today it is known as the Light's Golden Jubilee (http://www.prmuseum.org/blog/2015/8/24/lights-golden-jubilee). For him even in those days, when search engines didn't exist, words had the same potential to impact on human behavior. Thus, being a powerful instrument as well as a starting point on the search box, they require our special attention. To get inspiration and a real-time picture, visit www.google.com/trends/ where you can see what people all over the world are searching for. Furthermore, we can also read Bing Blog (http://www.bing.com/blogs/) for information about what people there are searching for. That information has to be used as a guideline, to avoid copying the most popular terms. In most cases the right keywords may not attract a huge number of visitors to your page, yet they are bound to involve the right type of potential guests. If you focus on people's needs and what words they will use to find your event, you will get the right visitors. Remember that your event is aimed at fulfilling

Google PageRank Prediction	Rank Checker
Predicts your future Google PageRank	Keyword position toll
http://www.seomastering.com/pagerank-prediction.php	http://smallseotools.com/keyword-position/
PageRank Checker	**Link Popularity**
Free tool to check Google PageRan, domain authority, global rank links and more.	Find how many links a domain has on the most common search engines.
http://checkpagerank.net/	http://www.iwebtool.com/link_popularity
Backlink Checker	**Search Engine Position**
Get complete detailed information about quality and quantity of backlinks pointing to your website.	Track your keyword rankings online
http://www.backlinkwatch.com/	https://serps.com/tools/rank-checker/
Multi-Rank Checker	**Reverse IP/Look-up**
Check multiple domains PageRank and ratings.	This tool will do a reverse IP lookup.
http://multipagerank.com/	http://mxtoolbox.com/ReverseLookup.aspx

Figure 3.7. Frequently Used Tools for Page Rank Checking

some need. Think about that need, what words they will use to find a solution, focus on those words and explain in the body text why your event is the best solution for their problem. To get a good idea of what the best keywords for your event are, you can try some of the available webmaster tools.

It is a good idea while searching for the most suitable keywords to invest a small amount of money in sample campaign in Google Adwords and/or Bing Adcenter. Be very careful and track down the impression and conversation rate for those words, which have 2–3000 clicks. These are bound to be your best keywords.

A well-known fact about searching is that the most popular words make up less than 30% of the searches performed on the web. The remaining 70% are related to searches for specific words or a combination of words, which are, in fact, a closer and more accurate representation of the needs of users. The full story of the "long tail" (Anderson,

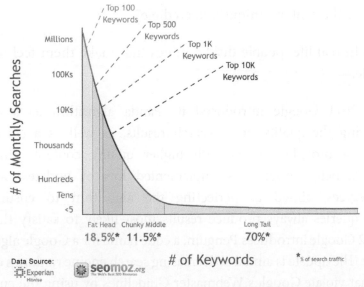

Figure 3.8. The Search Demand Curve (Source: Fishkin, 2009. Used with permission.).

2006) is something that has to be read, yet for the purposes of our study, it is important to point out that those who search for popular terms are not the ones who make purchases. They are just passing down information, while those who know what they need and want are the people who search for specific keywords.

EXPERIENCE ON THE WEBSITE

No one knows exactly the whole algorithms used by search engines. What is certain is that they supply users with the best possible results in their aim to meet users' needs. In doing so, they monitor how people interact and engage with your content, collect information, which they analyze, and offer the best answers. In order to make your site more attractive and useful to users, you have to find the answer to the following questions:

- Is the website easy to use and understand?
- Is the information relevant to what we promise to be?
- Is it accessible to modern browsers?
- Is the content unique and credible?

As in real life, people dislike things that make them feel uncomfortable.

In 2011 Google introduced its Panda algorithm as a way of improving the quality of its search results, as well as a convenient means of providing users with higher quality content. Since the initial launch, Google has implemented lots of updates of varying degrees, aimed at perfecting the algorithm to ensure that search queries always produce results most likely to satisfy the user. In 2012 Google introduces Penguin, a code name for a Google algorithm update. The update is aimed at decreasing search engine rankings of websites that violate Google's Webmaster Guidelines by using the currently declared black-hat SEO techniques involved in increasing artificially the ranking of a webpage by manipulating the number of links pointing to the page. It is obvious that Google is constantly changing its algorithm to guarantee its users the most precise answers and put a stop to all manipulation. The only thing to be claimed with certainty is that if you have useful, unique and relevant content, those changes will not affect you in the least.

Thus, what you should do is consider carefully what the intent of the search is, so as to be able to offer the best answer. With regard to events, there are a few important points worth mentioning:

1. People want to identify a local event, make a reservation, buy tickets, or register online.

2. People want to find your event on the web to get information, details or to share it.

3. People just want to get more information.

We have to work out our pages in a way to meet those needs step by step. We can respond to the needs of our users who, in fact, are our guests. We can offer them content in a variety of forms—text, audio, pictures, video, and charts—due to the popularity they have gained in multimedia communication nowadays. Last but not least, we have to make the whole content easy to share.

Links, links, links

In brief, link building is the process of establishing relevant, inbound links (incoming links) to your website, which help the latter achieve a higher ranking with the major search engines and drive targeted traffic to your site. The successful link-building campaign may not bring you fast results. In such a campaign the main goal is not simply to have links to your website but to build links from credible sources. Even though it is a key element, link building on its own cannot help your site to achieve its natural ranking. However, provided that it has become an incorporated part of your SEO approach, your site will gain the full benefits from links. Here are some suggestions for link building:

#1 Include links to and from websites that are popular in a city in which the event will take place, such as city guides.

#2 Use clickable links in the anchor text, link label, link text, or link title, and make the hyperlinks visible. The words contained in the anchor text can determine the ranking that the page will receive by search engines. Not all links have anchor texts because it may be obvious where the link will lead, due to the context in which it is used. Anchor texts normally remain below 60 characters.

#3 Choose carefully which websites you need to put links to; the more credible the site is, the better it is to have a link to it.

#4 Keep your content fresh. Practice proves that links in fresh content have a more significant impact on rankings, which means we have to aim constantly for links in new content.

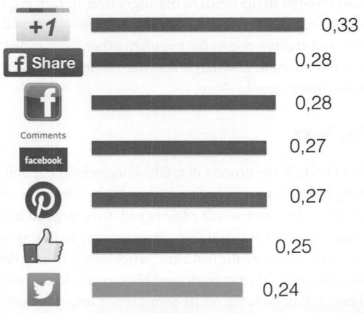

Figure 3.9. Social Media as Factors in SEO Rankings 2015 (Data from SmartInsights: http://www.smartinsights.com/search-engine-optimisation-seo/seo-strategy/seo-correlation-studies/attachment/seo-ranking-factors-2015).

#5 Aim for social shares. Here we can share links on our pages, profiles, etc., which may also stimulate users to share with their networks.

Although link building may seem very complicated, actually it is not. We can take some small steps to increase the number of links to our website. As long as we have created the best content on the topic, we will receive natural links from other websites, which strive to make the life of their own users easier. Firstly, we have to submit our website to directories and include the URL in our signature in every single e-mail we send. We can ask bloggers to link to us. In addition, we can create links by ourselves when we comment in forums, blogs, and in our user profile.

In planning our link-building strategy we have to work closely with our content creators and communication experts. The best way to gain links is to create a newsworthy content that people are eager to share. First and foremost, we must have the content available to create the easiest way for people to share instantly. If there is a person in your team who loves writing, creating a blog is a good idea for generating links.

All search engines are constantly trying to detect those who pay for links in order to diminish their significance, so it is not reasonable to invest money in that practice. On the contrary, invest in creativity and resources to create content that people will be willing to share.

Share this!

We are all familiar with that button. And we know that when we find something interesting, we want to share it with our networks. A more advanced way to increase shares is to add social share buttons to individual pieces of content on your website or blog. If we look carefully at the results of the research "SEO Ranking Factors—Rank Correlation 2013 for Google USA" (Buuteeq + Skift, 2014), we will find out how important social shares are. Of course, we must not forget that the research has laid its focus on Google, so G+ is on the top. Nevertheless, we have to be on the major networks and share links there. Although social media and networks will be discussed further in the chapter, it is important to point out, even at this stage, that we have to be where our potential guests are, impossible as it is to be everywhere at the same time. What's more, we have to be on G+, Facebook, Twitter, Pinterest, Youtube, and Flickr at least. The wider scope of our video content will enhance the potential and impact of our page rank.

We would like to draw your attention to a few guidelines for better social media presence in terms of SEO.

#1 Step. Optimize your profile in social media

Take advantage of any opportunity to write about your event, which means you need to fill in the "About" or "Information" sections of any social media platform. Use your keywords and phrases because they describe your event precisely. Moreover, they are also terms individuals would use to search for your event. Think globally but act locally. Although you may have heard this advice frequently, have you actually followed it? You should not forget to show your address, city, state and zip on your event page on Facebook. Always include links from your social channels back to your business's website (and links from your website to your social channels).

#2 Step. Optimize your social media content/updates

This means you should use your keywords once more in your posts, updates, and descriptions of photos and videos. Share all new content from your event website in all social channels. Sharing new content on Twitter is especially important because it helps Google to add your URL to its database. One more tip: It will help if you use your business's name in your social posts. In that way Google can easily associate the keywords you use with your business's name.

#3 Step. Optimize the opportunity to share

Shareability is all about what the post does to people: it affects them in such a way that they want to share with others. +1s, likes, repins, retweets, comments, etc., all improve the reliability of your links. The more share you get, the more inbound links are created. People share not only written content; it is just the opposite. "Content" doesn't always have to be as elaborate as a blog post or whitepaper; content can also refer to tweets or Facebook posts as well. By posting engaging social content, you're improving your SEO value. To review: Quality content leads to more shares, which is equivalent to more links, and they result in a higher SEO value. Simple, isn't it? In their study, "The Psychology of Sharing," The New York Times Insight Group (Brett, 2014) found that:

- 75% of the respondents say sharing helps them understand better and process news they're interested in

- 85% claim that the responses they get from posting and sharing on a social media site are an important source of thought

- 94% consider how helpful a link would be to another user before posting it

- 68% see sharing as a way to promote themselves and as a means to give other people a better idea of who they really are

- 73% say it helps them find people with common interests (Brett, 2014)

If social media and sharing is important to your event, you have to plan carefully your social content, keeping in mind why people share, give +1, retweet, etc.

If we do a thorough research into the type of content people share most often, we can come up with the top seven functions of highly shared post. Highly shareable posts do at least one of the following:

1. GIVE: Offers, discounts, deals or contests that everyone can benefit from, not just one sub-group of your friends.

2. ADVISE: Tips, especially about problems that everyone encounters; for example, how to get a job or how to beat the flu.

3. WARN: Warnings about dangers that could affect anyone.

4. AMUSE: Funny pictures and quotes, as long as they're not offensive to any group. Sometimes humor may not be quite as strong or edgy, it still has to appeal to a general audience.

5. INSPIRE: Inspirational quotes.

6. AMAZE: Amazing pictures or facts.

7. UNITE: A post that acts as a flag to carry and a way to brag to others about your membership in a group that's doing pretty darned good, thank you very much. (Carter and Marketto.com, 2014)

#4 Step. Share visuals

Pinterest is a great tool for link building and improving the keyword strategy for your event. When possible it is a good idea to use high-quality images on Pinterest, as Google prefers high-resolution images. Google indexes pages from websites with heavy traffic faster and higher. What's more, Pinterest is just one of those pages that you can add a link on to. However, do not forget to add keywords within the title of a Pinterest board and the board's description.

As a bonus you have up to 500 characters to describe an individual pin connected with your event. And here again you have to use your keywords. From an SEO perspective, Pinterest allows you to do a few things that other networks don't, so take advantage of it.

#5 Step. Spend time in Google+ every day

As Google is the number one search engine, we have to take into consideration its preferences. Sign up for a Google+ Business Page. Fill in as many fields as you can in the "About" section. Do not forget to use your keywords and phrases to describe your event. Google allows you to add customized links to your profile. Use this opportunity to link back to your website, blog, and additional social channels. Share all your new content with your Google+ page, which will increase your chances to appear within Google's SERP. If you are doing your event on a regular basis, your presence on the local listing is obligatory. According to Google, 97% of the consumers search for local businesses online. Google Places for Business gives you access to free tools that help your business get online, or to be found on Google Search and Maps, and get closer to your customers. It's the easiest way to manage your business across Google and it's free.

As social media and networks are quite new to all of us, we can— and do—make mistakes. However, the good news is that most of them can easily be corrected. So, if you have done a Facebook profile for your event, instead of Facebook Page, you can fix it (Lucey, 2014). If you have not added a customized Facebook URL, you can fix it (Pinkham, 2014). Check the name of all files you upload—text, pictures, and videos. Those names can be part of your SEO, so use keywords and/or phrases in their names (Secure a Vanity URL, 2014).

MEASURING

One of the greatest benefits of new technologies is that they give us unlimited opportunities for measuring what used to be immeasurable in the past. You can use the analytic data that is most convenient to your event's needs. In its review about 2014 Web Analytics Product Comparisons, Brian Lee says that, "Your customer has an itch, and your website needs to be able to scratch that itch. Knowing exactly how to do that is the key to getting visitors to take action on your site. You can use information gathered from web analytics to maximize revenues from pay-per-click advertising, improve click-through on email campaigns, or even increase your search engine ranking. The possibilities are endless" (Lee, 2014). There are free tools for measuring, which give you an accurate picture of what is happening and you have to find out why this is so. You can start by creating Google Alerts for all important keywords and phrases for your event. In fact, Google has a lot of tools, and most of them are free or have a free version: They include:

Google Alerts

Google Analytics

Google Insights for Search

Google Webmaster Tools

Google AdWords Keyword Tool

Google Trends

Google website Optimizer

DoubleClick Ad Planner

Google Page Speed

Merchant Center

Google Site Map

Google+

You can use tools from Yahoo!, such as Yahoo Web Analytics and Advertising Insights. It is important to point out that whatever can be measured can be improved. The focus lies on technologies and on the tools they offer us, yet there is a human being in front of every single computer. As humans, we tend to behave in the same way as we used to hundreds of years ago, which makes the following advice for better SEO by the "Father of Public Relations" quite relevant. The campaigns of Edward L. Bernays are still an example for those who want to change public opinion toward a cause, product, or event. Working long before the invention of the Internet, he had used some principles which can be applied to SEO even today. The essential lessons that Bernays teaches about link building can be summarized as follows:

Influencing the influencers. This means finding online influencers on the topic of the events and creating personal relationships with them. What is even more important is creating content, significant to them, which they will be willing to share with their networks. Those influencers have networks of followers who trust them, and that's the best way to spread information in our network society.

Help people express themselves by linking to you. The best way to do this is to make your content (event) a tool of self-expression. It is by far our greatest advantage that with the help of new technologies and a dose of creativity it's easy for people to express themselves by linking to our event. Walking in our potential guests' shoes, we can create numerous opportunities to offer them content—high quality, relevant, personalized, and delivered in context. PicBadges (www.picbadges.com) is just one among a bulk of possibilities at our disposal. It is a meteoric social platform, which offers people and businesses an engaging and viral way to reach their audience by creating and sharing ideas, activities, events, and interests on top of profile pictures on Facebook and Twitter. However, there are virtually no limits to the process of creation of different platforms as well as other ways to engage guests as early as possible.

SUMMARY AND CONCLUSION

Before coming to any conclusion about your SEO plans, try walking in the shoes of your guests. Try to learn as much as possible about them and what they will need. After that you have to focus on finding the best keywords and phrases for your event. Try to find out whether they work with either friends or friends of your friends. When you are ready, integrate those keywords on all possible occasions and start putting effort in creating the most relevant content for your event, which deserves to be shared.

DISCUSSION QUESTIONS

1. How do you make up your list of keywords and phrases?

2. How do you decide where to focus your link-building strategy?

3. How do you improve the shareability of your posts in social media?

TASK

Create a post for your profile on any social media, following the advice in this chapter. Note how people react to your posts. Share if there are differences in people's reaction on different social media platforms.

REFERENCES

Anderson, Chris (2006). The Long Tail: Why the Future of Business is Selling Less of More. New York, NY: Hyperion.

Apache Software Foundation. http://httpd.apache.org/docs/current/mod/mod_rewrite.html. Accessed 16.04.2015.

Brett, Brian., The Psychology of Sharing, http://nytmarketing.whsites.net/mediakit/pos/, accessed on 27.01.2014

Buuteeq + Skift. 2014. SEO Ranking Factors – Rank Correlation 2013 for Google USA, http://skift.com/2013/09/24/3-studies-show-social-media-drives-hotel-brand-loyalty/#1

Carter, Brian and Marketo, Contagious Content, http://www.marketo.com/_assets/uploads/Contagious-Content.pdf, accessed 27.01.2014

Chaffey, Dave (2015) "SEO Ranking Factors 2015." Smart Insights.

Fishkin, Rand (2009). "Illustrating the Long Tail." Moz.com. https://moz.com/blog/illustrating-the-long-tail. Accessed 01.01.2016.

Helicon Tech. 2015. http://www.helicontech.com/isapi_rewrite/ Accessed 16.04.2015.

LaFerney, David. 2007. A Complete Glossary of Essential SEO Jargon, http://moz.com/blog/smwc-and-other-essential-seo-jargon, July 26th, 2007, accessed 24.02.2014

Lee, Brian, Web Analytics Review, http://web-analytics-review.toptenreviews.com/, accessed 27.04.2014

Lucey, Blaise, How You Can Use Images for SEO, http://blogs.constantcontact.com/fresh-insights/why-you-should-optimize-images-for-seo/, accessed 24.02.2014

Moz. (2014a). http://www.viralseoservices.com/resources/seo-quote-rand-fishkin.html. Accessed 30.03.2015.

Moz. 2014b. Title Tag. https://moz.com/learn/seo/title-tag. Accessed 01.27.2014.

Pinkham, Ryan, Convert Your Facebook Profile to a Page, http://www.socialquickstarter.com/content/110-convert_your_facebook_profile_to_a_page, accessed 27.01.2014

Purcell, Kristen, Joanna Brenner, and Lee Rainie, Search Engine Use 2012, MARCH 9, 2012, http://www.pewinternet.org/2012/03/09/search-engine-use-2012/, accessed 24.02.2014

Secure A Vanity URL. 2014. http://www.socialquickstarter.com/content/33-secure_a_vanity_url, accessed 27.01.2014

Seomoz.org. 2014. http://d1avok0lzls2w.cloudfront.net/img_uploads/search-demand-curve(1).gif. Accessed 9/15/2015

SmartInsights: http://www.smartinsights.com/search-engine-optimisation-seo/seo-strategy/seo-correlation-studies/attachment/seo-ranking-factors-2015

TechTerms.com, 2014. Metadata. http://techterms.com/definition/metadata. Accessed 27.01.2014.

ADDITIONAL RESOURCES

The Beginner's Guide to SEO http://moz.com/beginners-guide-to-seo/how-people-interact-with-search-engines

Fleischner, Michael H., SEO Made Simple (Th ird Edition): Strategies for Dominating the World's Largest Search Engine.

SEO training program with tools, videos, a private member's forum, and so much more. http://www.seobook.com/

Williams, Andy, SEO 2014 & Beyond: Search Engine Optimization Will Never Be yhe Same Again!, Dec 6, 2013. Published by CreateSpace Independent Publishing Platform; 2.0. edition.

CHAPTER 4

Venue and Suppliers Search Technology

> We all use Google for search and agree that it is a great search engine in general. However, you type "a conference hotel in Washington DC," then it shows 12.4 million results. Where do event/meeting planners go from here? We definitely need an event-specific tool for meeting and event venue searching.

LEARNING OUTCOMES

As a result of reading this chapter, you will learn how to:

- Access an accurate and up-to-date global database of meeting and event venues

- Streamline site selection and sourcing process

- Research destinations, and find and compare meeting facilities faster and more easily than relying on traditional search engines

- Search venues and send request for proposals (RFPs)

- Source meeting space with ease and effectively use online venue search services

INTRODUCTION

Choosing a meeting/event venue is one of the most delicate decisions that event professionals face. A venue search was traditionally done in person—often known as MBWA (management by walking around), as being on site for the search used to be known as the best and only way. However, due to rising fuel costs and costs associated with staff time (planning, traveling, and attending), accommodation, and meals, the demand for an economical but effective venue/supplier search method has increased. An exponential increase in hardware capabilities is driving progress in all domains of IT and computing in meeting and event management.

Systems are more powerful than before and can share more information digitally. Also, advances in computers, CPUs (central processing units), and Internet broadband along with more information (video, image) of meeting venues are now easily shared via online sources (website, social networking sites [SNSs]).

ENVIRONMENTAL SCANNING

Most companies in today's event and meeting business are constantly strategizing and finding new ways to streamline operations to gain a competitive advantage. More and more event/meeting organizations are now using technology for their meetings and events, and the motivation usually comes down to "will it make a big enough difference?"

An industry-specific search engine database is a breakthrough for the meeting and event industry as it integrates technology with traditional event management tools to create more effective meeting/event planning. Today's meeting and event management experts rate demographics as one of the most significant challenges for event planners and suppliers. With a wide variety of pre-boomers, baby boomers, and Generations X, Y and Z, as well as a variety of cultural backgrounds, satisfying everyone is impossible. Event and meeting professionals are coping with demographic diversity by providing more options in technology, scheduling, meeting spaces, and even food and beverage choices.

Meeting Professionals International (MPI) (2014) found several changes in venue selection as well. Most notably, public perception—once a key factor in picking a city—has become one of the least considered elements for selecting an event destination. Just 14 percent of respondents say it plays into their thought process, while 42 percent name overall cost as the most important factor. In addition a strong move toward green (or environmentally sustainable) meetings is another force influencing event management, including venue and supplier search procedures. Excessive use of paper can be eliminated by utilizing an event management service such as venue/supplier search technology to communicate with venues. Site search software can be used for all facets of event management.

Other technologies that support green meetings include technologies that allow for stability and ease of business operations, for cost savings, for reduction of the carbon footprint, for adding value for clients, for innovation, for competitive advantage, and for staff motivation.

The computer integrated system (CIS) is the integration of multiple programs/applications to achieve a streamlined process of information and increased productivity as well as for reducing a redundancy of data and avoiding multiple entries of the same data.

INTERNET

Online communication and information dissemination over the Internet have significantly impacted the organization of meetings and business in general, all of which is made possible by the widespread availability of high-speed online access (initially ISDN [integrated services digital networks] and later fiber-optic cables).

The annual growth of broadband is between one percent and two percent for most countries, which will lead to even wider availability in the future. Fiber-optic cables, which have mostly been used for Internet backbone networks, are increasingly also used by private customers. In other words, Internet bandwidth will continue to grow, with consequences for many of the technologies relevant for meetings; video and multimedia will be more accessible and practical, enabling more effective forms of technology-based venue/suppliers search. Increasingly, the Internet has served as the backbone for a faster and media rich online venue search.

While improved Internet access was the most or second most demanded technology by both planners and suppliers, it was rated the 10th most difficult demand to meet by planners and the eighth most difficult by suppliers—representing one of the largest discrepancies between supply and demand.

In 2010, the penetration of broadband connection for businesses in 30 countries was more than 75 percent, and even in private households, many countries had a broadband penetration of between 60 percent and 80 percent, topped by Korea at 98 percent, according to data from the Organization for Economic Co-operation and Development (OECD, 2011).

Despite the conveniences of forgoing in-person site visits, the online venue search is still daunting, labor-intense work. For example, if an inexperienced event planner searches using general search engines (e.g. Google, Yahoo) for his event venue, it will surely overwhelm him. Further, if the search involves an unfamiliar destination, the challenge will be even bigger. One of the strongest focuses of technology providers over the past years revolved around facilitating this change. Venue search engines have popped up all over the place. Websites where event planners looked for a new destination or venue could find considerable information about potential candidates. Some companies automated the request for proposal (RFP) process. With the RFP, planners collectively prompted venue owners to reply to their demands. Undoubtedly this is one of the hottest areas of technology where innovation could result in multi-million dollar revenues.

BENEFITS OF VENUE AND SUPPLIER SEARCH TECHNOLOGY

Search technology, such as Cvent Supplier Network, can let meeting/event planners source, manage, and budget events using a single toolset. By storing key contacts, supplier notes, quote history, and other detailed reports in the same system, search technology makes the event planning process more efficient and transparent. The tool also manages cancelled space, preferred suppliers, and negotiated rates, and allows planners to find venues near precise locations with interactive maps. Some functions of this technology estimates capacities and converts between standard and metric units.

Easier site selection

Online venue search technology allows for the search of tens of thousands of hotels globally for meeting services from a single location. Event planners can quickly view property details and room specifications as well as surrounding attractions.

Easy and efficient RFP management

The RFP can save time by using established digital templates (eRFP). The eRFP allows meeting planners to send a request to multiple properties with one click. It also permits users to add a new venue or deselect an existing venue from the initial eRFP. The eRFP tool allows easy and quick comparisons of proposals from hotels side by side, allowing planners to gauge the most competitive bids simultaneously.

Contract management

This technology enhances efficiency in contract management by viewing details of contract rate history with a hotel, which allows for benchmarking against previously negotiated rates. It also allows an event organization to use established terms and conditions to existing eRFPs to prevent unwanted attrition fees.

For venues and suppliers

Venue staff can easily review, manage, and respond to eRFPs. An industry report says that this technology can respond up to 70% faster than before. It also saves time by reviewing all RFPs via a single page. Finally, it stores some pre-generated RFP context and answers that can save time to respond, which allows sales staff to spend more time evaluating RFPs for fast and accurate replies in order to win a bid. Table 4.1 provides a brief summary of advantages of online venue and supplier search applications for event organizations.

VENUE AND SUPPLIER SEARCH TECHNOLOGY COMPONENT AND PROCESS

Venue and supplier search technology provides the details that meeting and event professionals need to compare cities, venues, and suppliers,

Table 4.1. Comparison of Traditional Venue Search and Online Venue Search Application

Traditional	Related Activities	Online Venue Search Engine
Multiple e-mails or mails/ Fax	Sending RFPs	One-click for sending multiple RFPs
Mail delivery time + proposal preparation + return mail delivery time	Duration till receiving proposal	Within a business day
Multiple files or hard copies with different columns	Comparison of proposals	Easy to compare with standard format in electronic file
Multiple separate e-mails to notify a rejection	Reporting acceptance/ rejection	Easy with simple click in the system

helping planners identify the optimal location or suppliers providing services for their next event. Database systems are becoming a key information search and data storage tool. These software-based systems offer event planning tools that place every major venue, event space, and meeting space at your fingertips. Such technology sends multiple RFPs easily to multiple vendors by accessing an extensive and public database of hotels, restaurants, special event venues, and service providers as well as searching all meeting spaces by location and other criteria.

Components of venue and supplier search technology

1) Database

2) RFP

3) Filtering mechanism

4) Reporting

5) Corresponding module

Venue and supplier search technology process

A typical meeting/event venue and supplier search can follow various routes but can be summarized as follows:

Step1: Search

■ Enter the area you want to search, and the technology provides a list of venues in that area immediately.

■ Filter the local venues to find the room perfect for your function. Find event space based on relevant meeting criteria such as metro area, venue or service type, and total meeting space or chain affiliation.

Step 2: Creating RPFs

■ Creating a request for proposal (RFP) is simple with technology that is easy to follow and provides guides and time-saving features.

Figure 4.1. Diagram Digital Venue Search vs. Traditional Venue Search Process

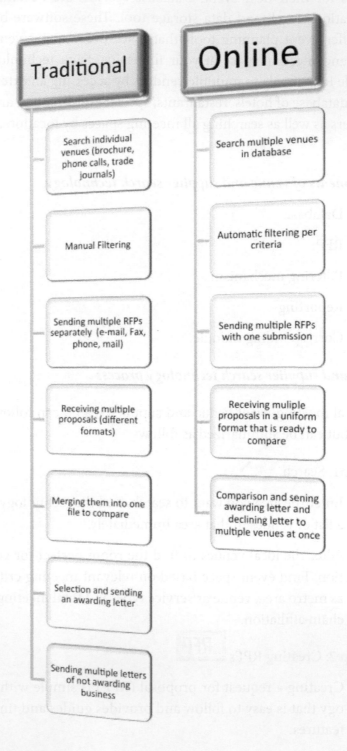

- Once planner and organization information is uploaded one time, future forms are pre-populated from existing account profile information

- Copy previous RFPs to eliminate data re-entry

- Attach key documents to the RFP

Step 3: Sending RFPs

- Select the venues that you want to send an inquiry to and the system will forward your details

- Save time by submitting RFPs directly to multiple venues at once, the National Sales Office, or both, enter in some contact details and any other quick details that will help with the venue response

Step 4: Receive bids and compare quotes from venues and service providers

- Compare bids side by side and select the best venues and service providers for your next event.

- Review multiple quotes side by side to analyze details such as availability, cost and meeting space.

- Notify suppliers of your decision to accept or decline their proposal

Step 5: Select the perfect venue

ONLINE VENUE AND SUPPLIER SEARCH SERVICE PROVIDERS

Cvent Supplier Network

cvent | *Supplier Network*

Cvent's Network provides event and meeting planners with streamlined site selection and sourcing tools. It helps planners to research

Figure. 4.2. Key Sections of Cvent Supplier Network (Adapted from Cvent Supplier Network 2014)

destinations and find and compare meeting facilities more easily and faster than traditional search engines. It is a complimentary tool for planners. See Figure 4.2 for a sample of a Cvent venue search tool.

Active MARKETPLACE

ACTIVE Marketplace from ACTIVE Network easily brings buyers together with tens of thousands of hotels worldwide—to make finding and securing meeting facilities online a snap.

It represents nearly 50% of the industry's top meeting services buyers on Business Travel News' 2012 Corporate Travel 100. And with the new ACTIVE MarketView 3.0, the intuitive cloud-based RFP management platform, hotel chains and properties can respond to eRFPs up to 70% faster than before (ACTIVE Marketplace, 2014).

Event planners can speed up and transform the process of sourcing for hotels and other meetings services via a single online location. ACTIVE Marketplace connects to, explores, and contracts with tens of thousands of properties around the world.

Venues can showcase their properties to qualified meeting planners and respond to incoming eRFPs in a timely fashion. It helps meeting venues to be connected with potential customers while easily reviewing, managing, and responding to their eRFPs via a single system.

Connect+ at Hilton Worldwide

The Hilton created Connect+, a dynamic collection of more than 100 of its largest hotels combined with their planning and event expertise and support. Every hotel in the Connect+ collection is in a destination city, with large-scale features including approximately 450+ rooms or 40,000 sq. ft. or more of meeting space. It believes in the power of bringing people together to share, engage, and connect for memorable events at exceptional destinations. Connect+ delivers results—return on investment in the form of both value and the valued connections that only face-to-face experiences can provide.

Further, virtual reality and 3D views of venues are easily accessible by a personal computer. These virtual site visits have drastically helped meeting/event planners save time during the initial stage of their search for a venue. Event venues' website contains image of the venue, decorations, and food and beverage options/price/photos. Thanks to advances in hardware and ICT (information and communications technology), now many event venues provide 365 degree panoramic views or virtual walk-throughs of the venue using virtual reality (VR) technology. Virtual reality has been a domain of mostly technical simulation. With immersion, 3D graphics and sound, the user experience simulates aspects of reality.

For meetings, virtual reality can replicate parts of an actual meeting venue and simulate interactions that are customary at face-to-face site inspections. Simulating meeting spaces online through graphic illusion simulates the face-to-face experience. Figure 4.3 shows an example from Hilton Chicago.

Venue Photos, Videos & Virtual Tours

The Portsmouth Ballroom is an elegant setting for your next event

Figure 4.3. Screenshot of VR Based Venue Inspection Tool (Source: iPLAN by NEWMARKET, used with permission)

SCREENSHOT

CIC APEX and Venue Selection/RFP Standardization

The meetings and events industry was realizing the need for standardization of processes due to the rapid developments in technology. In 2000 the CIC (Convention Industry Council) established a committee to develop accepted practices for the meetings and events

industry. This committee developed the Accepted Practices Exchange (APEX) that has been responsible for developing standardized forms and, most recently, standard reporting practices.

The benefits of implementing industry-wide accepted practices include:

- Time and cost savings

- Ease of communication and sharing of data

- Streamlined systems and processes

- Enhanced professionalism

- Superior results

APEX developed the standard venue/destination checklist that is widely used by many meeting event planners and venues to create their RFPs.

THE IMPACT OF APEX BY CIC ON VENUE/SUPPLIER SEARCH TECHNOLOGY

1. **RFP standardization**

   To develop recommended industry accepted practices for consistent and thorough requests for proposals (RFPs) that address core information and unique needs.

   CIC APEX Request for Proposals initiatives: http://www.conventionindustry.org/StandardsPractices/APEX/RequestsforProposals.aspx

   Manage Event RFPs using the new APEX RFP Workbook: http://www.conventionindustry.org/StandardsPractices/APEX/RFPWorkbook.aspx

   The APEX RFP Workbook offers planners a way to create and organize the most common RFPs associated with a meeting, convention, exhibition, or event in one convenient location. The new APEX RFP Workbook takes a streamlined approach, removing as much unnecessary or duplicative information as possible and providing simple, drop-down menus and auto-populating fields where possible.

Benefits of new APEX RFP Workbook for all meeting professionals:

- [ ] Save time by reducing retyping of basic event information

- [ ] Quickly create RFPs using drop-down menus and automatic calculations

- [ ] Produce clean, professional documents

- [ ] Stay organized with all major RFPs in one place

2. **CIC APEX meeting and site profiles initiatives**

The purpose of the initiatives were to develop recommended industry accepted practices for consistent and thorough profile formats for sites, as well as meetings, conventions, and other events, that include both core and unique information. After a great deal of research and discussion it was decided that the APEX meeting profile portion of the document is better suited to be contained within the request for proposals (RFPs) Panel rather than as a stand-alone panel (Convention Industry Council, 2014).

SUMMARY AND CONCLUSION

The meeting/event venue and supplier search technology is a combination of WWW, Wi-Fi, and cloud technology in the Meetings and Events Technology Golden Triangle.

The meetings and events industry has experienced dramatic change. Venue and supplier search technology provides the details meetings and event professionals need to compare venues and suppliers, helping planners to identify the optimal location or suppliers providing services for their next event. An exponential increase in hardware capabilities is driving progress in all domains of IT and computing in meeting

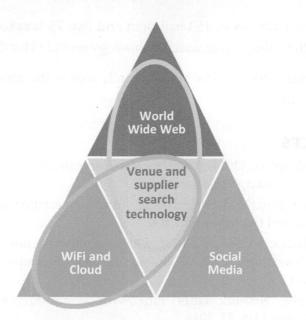

Figure 4.4. Meeting/Event Venue and Supplier Technology in the Meeting and Event Technology Golden Triangle

and event management. As database systems are growing as key information search and data storage, software based on database systems offers a useful tool for event planning by placing every major venue, event space, and meeting space at your fingertips.

DISCUSSION QUESTIONS

1. How does the APEX meeting event site profile and RFP initiative impact the development of meeting and event venue search technology?

2. What are the key benefits of using meeting/event venue search technology?

3. Describe the differences between successful meeting/event search technology business models and those that are not.

TASK

1. Develop a RFP for a 150 attendees for a one night, two-day association meeting in Washington DC. It requires two meeting

rooms (capacity of 150 ballroom and two 75 breakout rooms). Use the online venue search technology provided for this activity.

2. Discuss which online venue search tool is the most effective and why.

REFERENCES

ACTIVE Marketplace. (2014). http://www.activeevents.com/solutions/product/active-marketplace, accessed Jan. 11, 2014.

Connect + at Hilton.(2014). WorldWide, http://connectathiltonworldwide.com, accessed December 13, 2014.

Convention Industry Council. (2014). APEX, http://www.conventionindustry.org/standardspractices/APEX/MeetingSiteProfiles.aspx, accessed Jan. 5, 2014.

CVENT Supplier Network. (2014) http://www.cvent.com/en/supplier-network, accessed Jan. 31, 2014.

Hilton Chicago. (2014) http://www.hiltonchicagohotel.com/photo-gallery/360-tour/, accessed March. 5, 2014.

MPI (2014). Meetings Outlook, http://www.mpiweb.org/docs/default-source/research-and-reports/meetings-outlook-spring-web.pdf, accesses Feb. 11, 2014.

OECD (2011). OECD Broadband Portal, http://www.oecd.org/internet/oecd-broadbandportal.htm, accessed Feb. 11, 2014.

OTHER ONLINE VENUE SEARCH TOOLS

- http://superbook.eventmarketer.com/category/venues_properties/
- http://venuejar.com
- http://www.venuefinder.com
- http://www.venuemirror.com
- http://www.ultimatevenue.com/event-venue-finding.html
- http://www.functionvenuefinder.com.au
- http://connectathiltonworldwide.com/?WT.srch=1
- http://www.eventective.com

Part II

Meeting and Event Technology for Design, Planning, and Evaluation

CHAPTER 5

Meeting and Event Design Technology

> *"A designer is a planner with an aesthetic sense."*
>
> —*Bruno Munari,*
> *Italian artist, designer, and inventor*

LEARNING OUTCOMES

As a result of reading this chapter, you will learn how to:

- Apply event and meeting floor design technology

- Evaluate available meeting and event design technology

- Apply event and meeting floor plans for better attendee service before/during/after an event or meeting

- Forecast future trends in event and meeting design technology

INTRODUCTION

Today's event and meeting venues are getting larger and more technologically advanced. Their layout is not only getting efficient but also consistently being renovated to accommodate various requests from planners and attendees. For example Walter E. Washington Convention Center, opened in 2003 in Washington, DC, has 2.3 million-square-foot space of all sizes, from space for small groups and break-out meetings to events for 500 to 42,000 attendees. It includes a range of mixed-use exhibit spaces, 198,000 square feet of flexible meeting space with a total of 77 break-out rooms. Imagine you are the event planner at this venue and design all necessary logistics for keynote sessions, breakout sessions, an exhibition, an award banquet, a rooftop reception. There are many factors to consider for a layout of event/meeting room assignments, including room set-up, number of expected attendees, topics of session, time that it takes to go from one meeting room to another meeting room, number of doors to enter/exit, and so on. Planning for all this will certainly entail a tremendous amount of work and will require support from professional floor diagram design professionals and technology.

Not long ago, an official event floor diagram was drawn by a computer-aided design (CAD) specialist. CAD is very useful to create a professional scaled floor plan. It is a hard-copy-based professionally drawn blueprint or an electronic file that can only be accessed or modified using CAD software. A sample of a hard-copy blueprint printer can be seen in the Figure 5.1 below. As details of an event often can be changed until the day of event, consequent changes on the floor diagram had to be made via multiple back and forth communications between an event planner and the CAD professional. Unfortunately event/meeting/tradeshow planners were not able to do any revisions on their own for every need as it required the high-priced CAD software, a plot printer, and most of all, special expertise in CAD operation. Therefore any minor ad hoc changes or urgent needs of changes in the event floor often had to bypass the CAD company and were communicated using verbal or a non-scaled brief drawings of floor plans. Most of all, changes in exhibitor booth or session changes made on a master blueprint were not automatically updated to those who share a copied floor plan. It had to be done via multiple phone calls or e-mails to inform any changes made on the master blueprint.

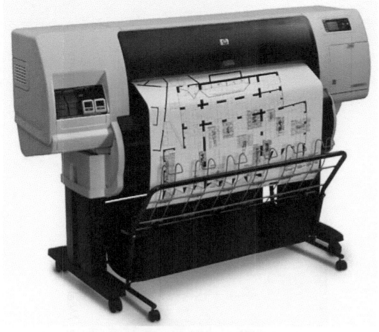

Figure 5.1. CAD Software Drawn Floor Plan on Printer

One of the top trends that has impacted the meetings and events industry is a greater utilization of technology, and new technology-based event floor design software came to the meeting/event market. These programs let event/meeting planners use pre-installed extensive interactive venue diagrams for easy drawing or for creating an original diagram if a venue is not in the database. As these diagrams are electronic files, it is easy for planners to create as many diagrams as needed for clients and to communicate with key event staff (e.g. venues' banquets and catering staff). It is also capable of rendering a 3D event diagram and printing, saving and sending it via e-mail to anyone who needs the diagram.

These event diagram design software programs are easy to use to create professional looking diagrams without extensive CAD training. These programs are personal computer-based and adopt user-friendly graphic interface applications, which means all operations are easy to see and perform, with most of functions using a simple click, drag, and place system. For an example of graphic user interface-based event diagram design software, see Figure 5.2.

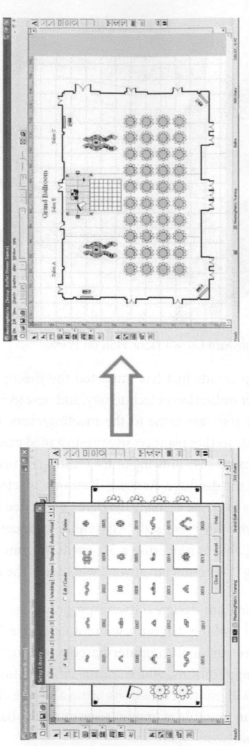

Figure 5.2. Pre-Designed Icons and a Sample Event Diagram (Source: MeetingMatrix, 2015; used with permission).

More and more event diagram design applications are going online so event/meeting planners and key stakeholders have full access to these software packages and diagrams as long as they have Internet access. As cloud computing becomes a standard in many IT systems, software does not need to be installed on computers but can be accessed by everyone involved via cloud computing, which is the practice of using a network of remote servers hosted on the Internet to store, manage, and process data.

Virtual reality venue tour (3D)

There are companies that offer 3D virtual reality that can create photo realistic tours from blueprints and interior design specifications. As a result of these programs the event venue sales professionals are able to show current and prospective clients exactly what their venue would look like when an event is hosted.

BENEFITS OF EVENT AND MEETING FLOOR DESIGN SOFTWARE

1. For planners:

   Event floor diagram software can facilitate better event planning coordination. Planners can e-mail event set-up diagrams directly to their clients for effective discussion on set-up and specifications. Planners also can print out diagrams to share with key stakeholders. Professionally drawn and scaled diagrams are shared with venues for accurate room layout as these diagrams reduce guesswork and unnecessary back and forth communications between planners and venues. When implementing event diagram design, technology can shorten the planning process.

   More and more event/trade show and meeting planners successfully create, manage, and market their events on diagram design tools on platforms that range from the desktop to the Internet to mobile devices.

Online-based event floor diagram design tools are used by most show organizers and general service contractors in the today's event world. These products include exclusively designed versions for general service contractors (e.g. GES, Freeman, etc.) and professional event/meeting/trade show organizers in the trade, corporate, consumers, and festivals and fairs industries. What is the most impressive about today's diagram design tools is that they can be seamlessly integrated with dimensionally accurate and rich revenue creating features.

2. For event and meeting venues:

Event diagram tools allow meeting/event venues to market and sell their event space online in a dynamic and engaging way. They provide a virtual tour of venues that their sales team can use onsite or remotely to engage and close event business. It's an online interactive experience that event planners use to discover a venue's capabilities. Venues can showcase their properties in

Table 5.1. Benefits of Online Event Floor Diagram for Planners and Venues

	Meeting/event/exhibition planners	Venues
Benefits	– Integrates event coordination with accurate and easy to update floor plan – Shortens planning process – Reduces unnecessary e-mails/calls due to client requests – Provides easy creation of guarantee professional event floor plan – Enables up-to-dated accurate real time exhibition inventories – Doesn't require advanced level of design skills and needs only a simple click-and-drag operation	– Presents an event venue in detailed interactive room diagrams/set-ups in 3D – Shows a visually rich walk through of a venue to prospectus – Markets and sells their event space online – Showcases its property in its surrounding environment

their surrounding environments or in interactive maps showing detailed interactive room diagrams/set-ups in 3D. Event venue sales representatives can use interactive event floor diagrams to provide their potential buyers a quick, engaging view of event spaces with visually rich tools.

Event diagram tools can also provide meeting venues' website visitors an ability to tour a facility online, check out room diagrams, and easily design event set-ups. Venues can utilize these tools for turn-key venue sales and event coordination with thorough engagement of customers. Table 5.1. shows the benefits of online event floor diagram for planners and venues.

MAJOR EVENT/TRADE SHOW DIAGRAM DESIGN SOFTWARE

1. EXPOCAD

EXPOCAD was the first graphical exposition management (GEM) software that was created specifically for show management and operations teams in 1999. It implemented the very first, real-time, Internet-based trade show floor design. It enables event/meeting/trade show managers immediate access and total control over inside and outside exposition space and offers the ability to share the floor plan with all staff from the first contact to the final invoice. It enhances streamlining of management of events by its automation and by maintaining all exhibitor content. Once floor plans are designed and approved by the general service contractor, sales or operations teams start sales activities by renting, un-renting, modifying, and reorganizing the floor plan space to reflect any changes. This type of online-based real-time trade show floor diagram design software allow the organizers to sell exhibition booths either live on-site or live online with accurate real time inventories.

Today over $1 billion (USD) of exhibition floor is managed by EXPOCAD annually (EXPOCAD, 2015). 80% of US based expos are designed using EXPOCAD software. EXPOCAD is used by major show contractors including Freeman Companies (global), Global Exposition Specialists (global), The Expo Group, Hargrove, and so on. Figure 5.3 shows several screens from EXPOCAD.

Features

- Accurate automated CAD-based show management tool.

- Integrates with the contractor's software on the backend administration side.

- Various display tools viewable from websites to all mobile devices.

- Floor plan functions.

- Financial tracking, invoicing, and reporting tools.

- Integrates with Microsoft Word, to create customized letters, reports, and invoices, as well as multiple-page contracts with sponsorships and more.

- Available in a desktop version and in the cloud.

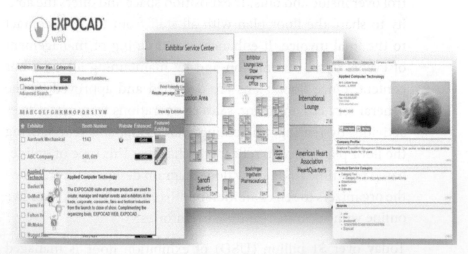

Figure 5.3. A Screen Shot of EXPOCAD (Source: ExpoCad) (http://new.expocad.com/page.cfm/Link=9/t=m/goSection=2_6)

■ Provides an analysis of any spreadsheet data using color-coded hatching that reveals the real patterns of attendee traffic on the show floor in color-coded animated heat map and video file formats.

2. MeetingMatrix

MeetingMatrix (MM) helps event planners evaluate, select, and work with venues. Its room diagramming software offers creation of accurate event diagrams and easy communication with clients and venues, resulting in streamlined collaboration and information-sharing during the events floor design process.

MM offers an extensive database of event/meeting venue floor plans. Currently, MM offers over 70,000 existing interactive venue diagrams stored in its database. When a selected venue doesn't have MM diagrams, the program allows planners to draw scaled floor plans using dimensions from the venue's capacity chart. MM has three levels of planner options: Express, Gold,

Figure 5.4. 3D Rendering of an Event Diagram (Source: MeetingMatrix, 2015) (https://www.flickr.com/photos/meetingmatrix/3717739674; used with permission from http://www.newmarketinc.com/.)

Premium+. Its introductory version, MeetingMatrix Express, is free. Figure 5.4 shows a 3D rendering of an event diagram.

Features

■ Easy to add set-up objects as you would with any other diagram.

■ Set-up Library: 80 pre-designed set-up configurations allow users to create and add frequently used set-ups for quick and easy planning.

■ StrataPictureIt: Inserts photos into the MM diagram, for example, a detailed photo of a table set-up or an ice sculpture to show customers and to guide the banquet team. Or show unique features for an event.

■ StrataStrata: Provides multi-layers that allow planners to show/hide details within diagrams. It allows various levels of set-ups depending on clients.

■ SeatingWizard: Generates event diagrams in seconds. All set-up components are accurate to the venue's particulars—tables, chairs, A/V equipment, staging, aisle spacing, and any unique items, enhancing the ability for perfect set-ups with all the venue's standards built in.

■ 3D-VR: View 2D diagrams in 3D virtual reality and allows virtual walk throughs.

MeetingMatrix iPlan is their newest cloud-based product. The iPlan Interactive Floor Plans platform (iPlan platform) provides the foundation to create cloud-based meeting solution modules. It utilizes the latest technology to optimize usability and performance. Each iPlan module is designed to be used independently or together for increased functionality. They can be accessed anywhere, at any time, using a desktop, laptop, or tablet. Meeting and event planning organizations have the flexibility to add the module(s) they need, when they need it. This is a further example of the pioneering contributions that MeetingMatrix has made to the meeting and event industry through cutting edge design innovations.

3. Social Tables

Social Tables is one of the leading event diagramming, seating and check-in services providers. Figures 5.5 and 5.6 depict a two- and three-dimensional event diagram and color rendering from Social Tables.

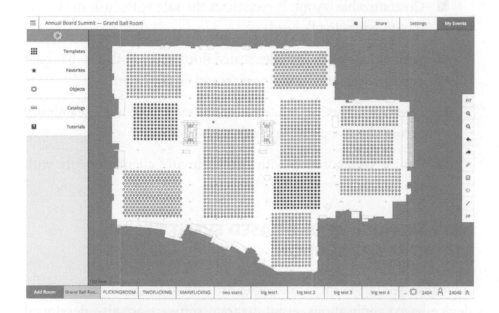

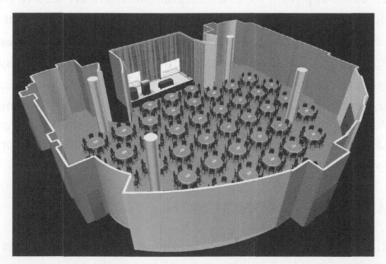

Figures 5.5 and 5.6. A 2D and 3D event diagram (Source: Social Tables, 2016; https://www.socialtables.com/learn/industry-resources; used with permission.)

Features

- 3D diagrams: Improves reality of venue review by highlighting the venue during the selection process with 3D diagrams. Impress clients with deliverables that show the value of your services.

- Customizable layout: Personalizes the sale with custom layouts that fit your client's needs.

- Scaled floor plan: Includes scaled floor plans in BEOs (banquet event orders) to sell when you're not in the room.

- Flexibility: Flexible solutions for any event and work seamlessly across any computer, tablet or mobile device.

- Online collaboration with clients: Invites clients to work with you online.

NEW TRENDS: MOBILE-BASED EVENT DESIGN DIAGRAM APPLICATION

As many meeting/event technologies are shifted to cloud- and mobile-device-based applications, event diagram software are also developed for mobile devices. Mobile event diagram apps let meeting/event planners easily and dynamically review venues on mobile devices (e.g, iPhone or iPad). These mobile apps provide interactive floor diagrams so planners can evaluate and preview event spaces and other features of the venue via easy-to-carry mobile devices. PlanView.mobile™ is an iPhone application that gives meeting planners and guests mobile access to venue floor plans and room diagrams. It allows easy, instant access to a venue to planners with an iPhone.

SUMMARY AND CONCLUSION

Creating easy and accurate event and meeting floor design was what meeting/event/trade show planners sought to help them execute effective meeting and event logistics design and boost exhibition sales by accessing real-time inventory. Today's meeting/event design tools are operated at the personal computer level and designed with user-friendly graphic interface applications so meeting/event planners can operate them easily with a basic skill set. Today, these tools are moving to a cloud-based mobile application format, which provides meeting/event planners mobility on top of easy and up-to-date event diagram sharing.

DISCUSSION QUESTIONS

1. What are the advantages of adopting event design programs/software compared to CAD?

2. What functions of event design software are effective with the 21st event/meeting technology trend (e.g. cloud computing)?

3. Please discuss advantages of utilizing online exhibition booth management to show organizers and exhibitors.

TASK

Create an event floor plan of various types of event and meetings (e.g. award banquet, general session, networking reception) using the discussed event diagram design tools. Then discuss strengths and weaknesses of each application.

REFERENCES

EXPOCAD (2015). http://www.expocad.com. Accessed April 16, 2015. http://new.expocad.com/page.cfm/Link=9/t=m/goSection=2_6. Accessed April 16, 2015.

MeetingMatrix (2015). https://customers.meetingmatrix.com/Public/Training.aspx. Accessed April 16, 2015.

Social Tables (2016). Accessed: https://www.socialtables.com/learn/industry-resources.

CHAPTER 6

Meeting and Event Administration Technology Solutions (Intranet and Other Internal Solutions)

LEARNING OUTCOMES

As a result of reading this chapter, you will learn how to:

- Define the role of technology in meeting/event administration management

- Identify current and forecast future meeting and event administration technology needs for your organization

- Select the appropriate meeting/event administrative technology for effective meeting and event management

- Explain the importance of information management and secure data management

- Evaluate and analyze meeting and event administrative technology options

INTRODUCTION

In most cases when we choose a carrier in the field of events, we dream about the creative part of the work, not about the administrative duties surrounding each project. In fact it is impossible to organize an event without administrative work. We are happy to live in times when technology can make that administrative much easier, faster, and, thus, more enjoyable. This type of work can also be creative.

In short the main goal of administration is to organize people and other resources efficiently and to direct activities toward common goals and objectives. If we go back in time, we can find that the word *administration* is derived from the Middle English word *administracioun*, which came from the French *administration*, itself derived from the Latin *administratio*—a compounding of *ad* ("to") and *ministrare* ("give service"). Hence, administration is "to give service". If we have that meaning in mind, we can accept that giving services can be a fascinating job and full of great ideas and challenging tasks.

According to *The International Dictionary of Event Management*, administration is "the group that determines budget, staffing required (organization chart), and flow of communications, as well as develops the timetable and production schedule for and event" (Goldblatt and Nelson, 2001).

As you see, good administration is an integral part of any event. Planning the event would be much more difficult and create more pressure without good administration. The team or the person responsible for the event administration, in fact, supports the planning process, helps in the implementation of the event, and assists in the evaluation of the event. It is not exaggerated to say that the administration of events plays a pivotal role in the success or failure of any event today.

Of course, not each event team has the chance to have a dedicated event administrator with special skills and knowledge of the specific field. But we have to keep in mind that that role is pivotal and should be executed by a professional. It is very common to see the use of event administration and event management as synonyms. In fact they are not. We have to know that there are significant differences between the duties of both functions. As event management is the clearer one, we

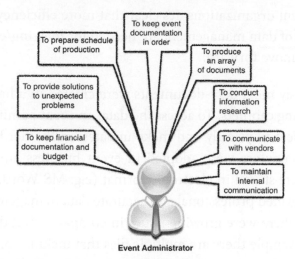

Figure 6.1. The Duties of the Event Administrator

will focus here on the event administration. The event administrator has to support the event from the beginning of the planning process by keeping documentation and tasks in perfect condition. Nowadays the event administrator needs a combination of administrative and IT skills to successfully complete tasks. We all know that each event entails a huge amount of documentation, which is in fact the history of the event. The event administrator is the person responsible for the task and at the same time she or he has to be able to conduct research, identify necessities, and keep in contact with vendors. Figure 6.1 shows the duties of the event administrator.

As you see the event administrator has many duties, and they are very different. This role is very important if we want to a produce great event.

DATABASE MANAGEMENT SYSTEM

Data is one of the most important assets a business has, but the rapid growth of data can make it increasingly difficult to manage. Today's business administrator handles a large amount data; therefore, effective and accurate data management is a critical part of great administration work. The converging trends of big data, modern apps, cloud, and bring-your-own-device (BYOD) represent challenges to many meeting/event organizations. It's clear that more efficiency and tighter automation of data management are needed for meeting/event administration to move forward.

Previously most data-documents were created in a hard copy and stored in filing cabinets. To access the data, an event administrator had to access them physically and make updates on the hard copy. Later, personal computers were adopted in event business administration, and data were stored in electronic format (e.g. MS Word, Excel, etc.). While it provided professional and accurate data management at a personal level, there were growing issues in company-level data management. For example there are multiple files that include common fields, but updating the information in one file system did not automatically update the same field in other files that were managed by other staff or departments.

The following section discusses the disadvantages of the old database management system, known as a traditional file system, and solutions to those problems with a contemporary file system (enterprise system).

Problems of traditional database management system include:

1. Data redundancy and inconsistency: Data redundancy is the presence of duplicate data in multiple data files so that the same data are stored in more than one place or location. Data inconsistency occurs when the same attribute may have different values.

2. Program-data dependence: The coupling of data stored in files and the specific programs required updating and maintaining those files such that changes in programs require changes to the data.

3. Lack of flexibility: The traditional system cannot deliver ad hoc reports or respond to unanticipated information requirements in a timely fashion.

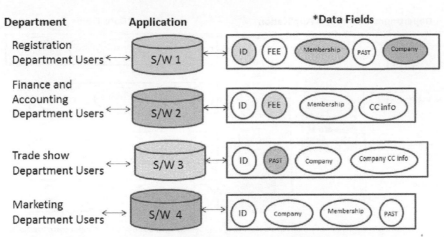

*Different colors of a same data field means their values are not identical

Figure 6.2. Traditional Information System of an Event Organization (Source: Rainer, Watson, and Prince, 2013)

4. Poor security: Because there is little control or management of data, management will have NO knowledge of who is accessing or even making changes to the organization's data.

5. Lack of data sharing and availability: Information cannot flow freely across different functional areas or different parts of the organization. Users find different values for the same piece of information in different systems, and hence they may not use these systems because they cannot trust the accuracy of the data. Figure 6.2 illustrates a traditional system with these issues.

Solution: A contemporary file system (enterprise database management system)

A contemporary file management system was developed to address the before-mentioned issues with the traditional system. Software now permits firms to centralize data, manage the data efficiently, and provide access to the stored data by an application program. It also acts as an interface between different application programs and physical data files As a result contemporary file management systems solve many problems encountered with the traditional data file approach.

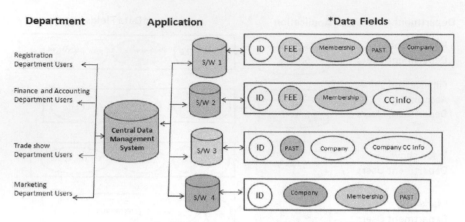

*Same colors of a same data field means their values are identical

Figure 6.3. Contemporary Information System of Event Organization

Contemporary data management systems have following advantages:

1. More uniform organization: Similar processes and information structure will improve administration management among various event meeting staff and suppliers.

2. More efficient operations: Contemporary file management can support efficient meeting/event administration by integration of multiple file management systems.

3. Customer-driven business processes: A contemporary system allows meeting/event administration system to connect customer input resulting in better meeting/event customer service.

4. Unified platform: The contemporary system used unified technology for various meeting/event administration systems.

The DBMS (database management systems) solved many headaches of event administration work-related data management; however, as data grows, it requires more space to save those electronic data, typically a server to save data. Servers require hardware, space to install them, and skilled IT professionals to manage them. Many meeting/event management firms are small businesses with limited staff and no designated IT specialist on hand most of time. Therefore the new wave of technology to change the game of database management has arrived to most business sectors, which is cloud computing.

ADMINISTRATION IN THE CLOUD
What Is Cloud?

The idea of cloud computing is surprisingly not new; it was first launched in 2000. Oracle's Larry Ellison launched the New Internet Computer (NIC) company to lead the industry forward to that goal. His concept was simple: the required equipment is simply a personal computer with just a processor, a keyboard, and a monitor. The computer does not require a hard drive or a CD/DVD player. The computer connects via the Internet to a remote supercomputer, which would host all necessary programs and files. The idea, however, was too advanced at the time, and there was a shortage of fast broadband in the U.S. back then. As a result the NIC closed in 2003.

Many business organizations try their best to make sure all of their administrative employees have the right hardware and software they need to do their jobs. This includes not only hardware (computer) but also software to give employees the tools they need. Cloud computing is the delivery of computing as a service rather than a product, whereby shared resources, software, and information are provided to computers and other devices as a utility (like the electricity grid) over a network (typically the Internet). Instead of installing software for each computer, one application allows workers to log into a web-based service that hosts all the programs the employees need to perform their jobs. The needed software is stored in remote machines owned by a company that runs needed software and data-management programs. With the fast advancement of faster Internet with broadband at work and at home, cloud computing has grown in popularity. Cloud computing eliminates installing software on the computer of each employee. This decreases demand for hardware and software on the end user's level. As long as an end user has interface software (often a web browser with Internet access) to access the cloud computing system, he/she can access the system that is managed by a remote company with 24/7 technical maintenance from anywhere as long as there is Internet access.

CLOUD SERVICE PROVIDERS

Microsoft Cloud Services

Microsoft delivers the broadest array and depth of cloud computing solutions to businesses. Its key areas are

- Cloud computing platform: It provides a unified platform, which is called Microsoft Azure, which allows an open and flexible cloud computing platform that enables users to quickly build applications using any language, tool, or framework, and to deploy and manage apps across a global network of Microsoft-managed data centers. Figure 6.4 shows Microsoft's cloud system.

- Cloud computing productivity suite: Office 365 offers productivity experience, spanning business intelligence, unified communications, e-mail, collaboration, and social capabilities.

- A private cloud solution: A suite of business solutions, Microsoft Dynamics helps organizations grow while managing costs.

Figure 6.4. MS Cloud

Google Cloud Services

If you have a Google account, which is nearly a must in today's online environment, you are certainly familiar with Google Cloud (https://cloud.google.com/). Google Cloud Platform is a portfolio of cloud computing products by Google, which is offering hosting on the same supporting infrastructure that Google uses internally for end-user products like Google Search and Youtube. One of the unique benefits is that when using Google Cloud, you have access to other Google infrastructures. As with all others cloud services, Google cloud offers the following benefits:

■ Secure and safe: Your data is protected through redundant storage at multiple physical locations.

■ Flexible access.

■ Competitive and flexible pricing: You pay for only for what you use.

One feature that can be very useful for event administrators is Google Cloud Print. This is a new technology that connects your printers to the web. Using it, you can make your home and work printers available to you and anyone you choose from the applications you use every day. Google Cloud Print works on your phone, tablet, Chromebook, PC, and any other web-connected device you want to print from.

Apple's Cloud Service

Apple's cloud storage product, iCloud, is designed to work seamlessly with all your Apple devices connected to the Internet. For example, you can upload photos from your iPhone and access them from your MacBook, upload music from your MacBook to listen to from your iPod Touch, or upload an important document from your Mac desktop to access from your iPad when you're on the go. iCloud's biggest advantage is that it's integrated into Apple software. At the same time, that can be a disadvantage if most of your team are not on Apple devices.

iCloud's features give you access to your data, from important contacts to fun photos, anywhere you're connected to the Internet. You can authorize up to ten devices to access and use iCloud with your Apple ID. This is a leap beyond the iTunes Store authorization, which is limited to five devices. Plus, iCloud authorization extends beyond iTunes to touch all apps capable of connecting and to use iCloud from that device.

Perhaps the biggest advantage of the iCloud service is how you can use it to back up and store data on your Apple iOS devices. iCloud is capable of taking daily backups of your iPhone, iPad, or iPod Touch when it's connected to the Internet using Wi-Fi. These aren't full backups, which would include all the data stored on each device. Instead, these are partial backups that store only the data you've changed on the device.

SERVICES FOR ONLINE STORAGE

Google Drive

Google Drive is a cloud storage service that allows you to store your documents, photos, videos, and more online in one place. From Drive, you can also access Google Docs, where you can create, share, and collaborate on documents, spreadsheets, presentations, and more from anywhere while online. From an administrator's point of view, Docs is a very useful service, because you can

- Create and edit documents
- Share your documents
- Revert to an older version of a document
- Merge PDF files
- Export documents into Microsoft Word, OpenOffice, PDF, and other formats
- Export files as Microsoft PowerPoint, PDF, JPG, and other formats to create presentations
- Create a blank spreadsheet that can be exported as Microsoft Excel, OpenOffice, PDF, CSV, and other formats

■ Create forms that can be filled out online. Forms can be exported to CSV files

In other words you can do you many jobs by using Google Docs and Google Drive. You need to have constant online access; otherwise your documents will not be available.

Another advantage of Docs in the field of event management and administration is the opportunity for people to collaborate when working on one document. You can decide which documents to share with whom, and you will see each person's changes and comments. Working together has never been so easy. Many event professionals use the option to create forms at the different points of the event process. It could be used to create different forms, such as

■ Preliminary questionnaires for the event

■ Registration forms for speakers

■ Registration forms for guests

■ Questionnaires for event evaluation

Google Docs also offers ready-to-use templates for event management, planning, budgeting, and more. It is possible to adapt the templates to your specific needs.

Microsoft Hybrid Cloud Storage

Microsoft is the pioneer of automated administrative technology. Every administrative employee with a computer spends a lot of time acquiring data and then trying to find a way to store it. MS Hybrid Cloud Storage integrates on-premises storage systems with cloud storage services. Its Windows Azure Backup Service provides a way for users to automate their nightly backup processes using Windows Azure Storage as the location for storing that backup data. This means that data no longer has to occupy on-premises storage, which frees storage administrators from the time-consuming and error-prone tasks of running and managing backups.

Advantages of MS cloud storage include:

1. Convenience: If you store your data on a cloud storage system, you'll be able to get to that data from any location where you have Internet access.

2. Flexibility: With the right storage system, you could also allow other people to access the data, turning a personal project into a collaborative effort.

The MS Office package provides various tools for administrative tasks (word processing, spreadsheet, e-mail management, etc.). MS Office is the most widely used administrative package in the world. It includes the programs Word, Excel, PowerPoint, Access, Outlook, and OneNote. Other MS desktop applications included in Microsoft Office suite include Microsoft InfoPath, Microsoft Publisher, Microsoft Project, Microsoft Visio, and SharePoint sites. It also offers clouding computing-based MS Office Online, which is a suite of web-based versions of Microsoft Word, Excel, OneNote, and PowerPoint.

Other Alternatives

Depending on the needs of your event team, there are other tools as well.

- Zoho Docs: This online document management program offers a free edition that gives users 1 GB of free space. Organizations with a broad diversity of operating systems, browsers, and mobile devices will appreciate Zoho Docs' cross-platform consistency. The main features are online file storage, secure file sharing, online chat and collaboration, and multi-level folders. It also integrates Zoho Writer, Sheet & Show, online workspace, reviewing and tagging, search, check-in/check-out control, and document version control.

  Zoho Docs supports the following devices: iPhone-iPad, Linux, Mac, web-based Windows. According to the online community, if the size of 1 GB is not an obstacle, try Zoho

Docs, especially if your team is using many different devices and platforms.

- Dropbox: This is an online sync and file sharing application. One of the biggest advantages is that Dropbox allows you to sync your files online and across your computers automatically. You can use 2 GB of online storage for free. The main features are mobile access and military-grade encryption methods. Devices that are supported include Android, iPhone-iPad, Linux, Mac, RIM-BlackBerry, web-based Windows, and Windows Phone.

- Evernote: This service helps users capture a note in any format (handwritten or text, web clip of a product review for reference, photo of a receipt, audio file) and have it be accessible and searchable on virtually any laptop, mobile device, or on the web. Also it lets multiple users collaborate on shared notebooks that can be instantly updated and accessed from the web, desktop, or mobile device. The free option includes 60 MB/month of new notes, unlimited total storage, and syncing across all your devices.

- One Drive: This is software from Microsoft. It helps to easily store and share photos, videos, documents, and more. The free option offers 7 GB free storage. Among the others key features of OneDrive are that it adapts video playback to your Internet speed, works seamlessly with Microsoft Office, creates surveys, collects responses and exports to Excel, browses animated GIFs, searches for text in your photo, uploads multimedia from mobile devices, has an automatic camera roll backup, posts to Facebook, sharse captions and geotags, etc. If many of your team are using Microsoft products, this is an option worthy of consideration.

- Hightail: This service will help you to send and share files and folders with anyone. With Hightail you can send files up to 10 GB from your computer or mobile device. You can also sign documents on Hightail and return them right away. You have the option to protect your files with a password and to ask for identity verification. If, in your work, you have to send and collaborate on or approve large files, Hightail is an option you have to consider. The free subscription offers you 2 GB of storage, shares files up to 250 MB, allows e-signatures, secures delivery, encrypts data, and provides mobile and desktop app access.

■ SugarSync: This is not a free service, but you can try it for 30 days for free. The cheapest plan is starts at $4.99 per month. With SugarSync you can sync all your files to all your computers and web-enabled mobile phones instantly, and you can also store and back-up your files in secure servers and access them from any web browser, anywhere, anytime. You can also share these files with your colleagues.

These are few of the products you can use for your administrative needs. At the time you read this book, we are sure that new options will be available. It is not possible to cover everything here. The most important thing is to carefully investigate the needs of your team and the devices they use and to find the best solutions for your needs.

In addition to working with your team, you may consider options to share content with your guests. In this case a good starting point is to learn more about the document sharing services available online. They can help you to get quick and easy access to the needed documents for all your guests on any devices. Take time and review some of the following options:

■ Yudu is a free service that allows you to upload PDF, Word, Excel, and Powerpoint files to create online magazines.

■ Issuu offers the option to display your documents with page-turning effects. You can embed your documents with those effects for your blog or website.

- DocStoc and Scribd are similar services. You can upload documents created with your preferred word processing program and share them online.

- Youblisher is a PDF publishing service that turns your PDFs into online magazines complete with page-turning effects, which can be embedded into your blog or website.

As mobile access to the Internet is growing, we will see in the future more and more apps to aid in administrative work. Ideally, they can help you to do your jobs from any device wherever you are located. Those apps can be used for the following tasks:

- Signing documents from mobile devices without paper

- Recording meetings

- Sending reminders for tasks

- Translation in different languages

- Note-taking, etc.

Be aware of the new products and choose those that will help you in your work.

SECURITY OF ADMINISTRATIVE INFORMATION IN THE 21ST CENTURY

Considering the wide adoption of cloud computing and database management, system security of information is increasingly critical. As mentioned, in the cloud system the cloud-based administrative management system decreases the probability of information stealing at the office level. The remote centralized data centers provide high security and 24/7 monitoring of any unauthorized attempts of system intrusion.

While key security management will be performed by trained security professionals, it is still important to have a good understanding of basic principles and mechanisms that have been successfully used by practitioners in actual products and systems.

PREVENTION, DETECTION AND TOLERANCE

The objective of data security can be approached in two distinct and mutually compassionate ways.

■ Prevention: It ensures that security breaches cannot occur. Access control is performed by the system to examine every action and check its compliance with the security policy before allowing access to occur.

■ Detection aims to detect any security breach after it occurs. The auditing process uses histories of recorded activity in the system to detect security breaches after the fact.

While prevention is the more fundamental technique, every information system employs various combinations of these two techniques.

BASIC SECURITY CONCEPTS

The objective of data security can be divided into three distinct but interrelated areas, as follows:

■ Integrity: This is concerned with improper modification of information or processes.

■ Secrecy: This is concerned with improper disclosure of information. The terms confidentiality or nondisclosure are synonyms for secrecy.

■ Availability (denial of service): This is concerned with improper denial of access to information.

These three objectives arise in almost every information system, including meeting/event administration database management systems (DBMS). For example, in the case of a payroll system, secrecy is concerned with preserving the confidentiality of salaries; integrity is concerned with preventing an employee from changing his or her salary; and availability is concerned with ensuring that the paychecks are printed on time as required by law.

The relative importance of these three objectives varies among different DBMS. For example most commercial DBMS, including meeting and event organizations' DBMS, place higher importance on integrity as it handles customers' both personal and financial information.

SECURITY POLICY

Security policies are established to clarify the before-mentioned three security objectives. A statement of security policy mainly consists of defining improper handling of the system. In general, a security policy is mainly determined within an organization rather than imposed by outside, but the proper handling of information is sometimes defined and mandated by law. For instance, legal and professional requirements apply to sensitive personal information (such as medical information) about individuals.

CONCLUSION

At the end of this technological chapter, we want to mention that we—as event specialists—are not in a position to solve one the biggest problem in 21th century: the security of information. Digitalization has made it much easier to reach a global audience, but it has also given rise to greater security dangers. There are some important steps we can take to help reduce security risks. These include:

- Reviewing carefully to whom we send e-mails. Sometimes it happens that we write the first name, and the computer finishes the e-mail address, and it may be that our intended recipient is is not the person by the computer.

- Sharing information only with people we trust.

■ Signing out from important apps on mobile devices when we finish using them. If the device gets in other people's hands, they will not be able to access our information.

When using mobile devices you have to keep in mind the following security measuress:

■ Use security software

■ Avoid phishing e-mails

■ Be wise about Wi-Fi

■ Lock up your laptop

■ Read privacy policies

The most underrated threat to online security is, according to Duncan Unwin, Chief Information Security Officer at *Your Digital File* (2014): *yourself*. He says: "It's security from your own mistakes. Being secure online should mean not just protection from hackers but it should also mean protection from someone making a mistake with their own data management, for instance losing passwords or private keys."

DISCUSSION QUESTIONS

1. How does digitalization change your personal documentation style?

2. What are the key features you should look for in choosing a cloud platform?

3. What are the advantages of cloud computing for meeting/event administration management?

4. What are the key security concepts that are emphasized in a meeting/event management database management system?

TASK

Evaluate the level of penetration of online services for collaboration in a project in which you take part.

1. List which products you use. Which of them are more effective and why? Create a list of criteria for choosing online platforms for collaboration.

2. Research what types of event/meeting administrative systems are used from your work sites. Are they traditional or contemporary data management system?

3. Review what type of security approach is adopted in your work site and who is responsible for information security in your work site.

REFERENCES

Goldblatt, Joe, and Kathleen S. Nelson (eds.). 2001. *The International Dictionary of Event Management,* 2nd edition. Wiley.

Rainer, R. Kelly, Hugh J. Watson, and Brad Prince, 2013. *Management Information Systems,* 2nd edition, Wiley.

TopTen Reviews.com Cloud-services-review.toptenreviews.com 10.06.2014). http://www.toptenreviews.com/.

Your Digital File, 2014. "Expert advice for online storage security. https://www.yourdigitalfile.com/expert-advice-online-storage-security/ Accessed 11.30.2015.

ADDITIONAL RESOURCES

Microsoft Cloud computing: http://www.microsoft.com/enterprise/industry/retail/hospitality-and-travel/default.aspx#fbid=EZgscI4kOb7 – Reduce carbon and overhead in the cloud.

Weinman, Joe, Cloudonomics (2012). www.getapp.com – a place where you can discover & compare more than 5,500 business apps+ Website: The Business Value of Cloud Computing, Wiley.

CHAPTER 7

Virtual Meetings and Events

百闻不如一见

"Hearing a hundred times is not as effective as seeing once."

Old Chinese proverb

LEARNING OUTCOMES

In this chapter, the current status of virtual meeting technology (VMT) will be explored to better understand the rapid development of this field of study.

As a result of reading this chapter, you will learn how to:

■ Understand the current and future impact of virtual meetings and events within the meeting and event industry

■ Understand key dimensions of rich media and the historical advancement of virtual meetings and events

■ Select the appropriate VMT solutions for meetings and events

■ Determine the appropriate levels of training that will be required to support these new technologies

- Understand various types of virtual meeting/event design and set-ups.

- Adapt new technologies for your specific virtual meetings and events.

INTRODUCTION

Technology continues to open up new ways to collaborate, share, engage, and discover. The way that people work has never been more dynamic. Meeting attendees and planners are sharing information quicker and getting feedback faster. As the mentioned ancient Chinese proverb suggests, visual is the most effective way of experiential learning and networking—which are two main goals of any meeting. The proverb *"Hearing a hundred times is not as good as seeing once"* can be explained by the media richness theory (MRT), which argues that face-to-face communication, a typical synchronous communication, is a rich medium and that it reduces uncertainty as it is a real-time visual communication. Kock (2005) argued that passing on and observing facial expressions and body language contributed to psychological arousal. Therefore, synchronous communication is considered as the most effective channel of communication. The Robert and Dennis's (2005) cognitive model of media choice (MMC) suggests that synchronous communication increases an individual's motivation and helps one to process messages. MMC also argues that high synchronicity increases motivation and decreases ambiguity. Specifically, remotely located learners are more easily distracted.

Synchronous communication is assumed to be more natural to all parties involved since it resembles face-to-face interaction. Kock's media naturalness hypothesis (MNH) (2005) assumes that natural communication involves co-location, synchronicity, and the ability to convey and observe facial expression, body language, and speech. Therefore, rich context of face-to-face communication is highly considered as an effective class communication and content delivery. Hrastinski (2008) introduced the concepts of personal participation and cognitive participation to describe the dimensions of participa-

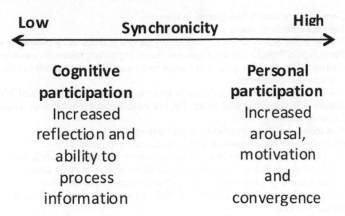

Figure 7.1. The Concepts of Cognitive and Personal Participation (Kock, 2001).

tion that are supported by synchronous and asynchronous communication (see Figure 7.1). An increase in the degree of synchronicity improves personal participation while a decrease allows more cognitive participation.

Innovative Internet communication technology (ICT) applications coupled with improved videoconferencing capabilities have led to the proliferation of VMT in recent years. There have been significant advancements in meetings and events through virtual/distributed digital technologies, such as teleconferencing, web conferencing, and Skype or Facetime. While many think virtual meetings and events are a very recent invention, there has been a long history of virtual meetings and events, as can be seen in Figure 7.2.

WHAT IS A VIRTUAL EVENT/MEETING?

While there is no universal definition of virtual events/meetings, this chapter frames the concept as an online event occurring at a specific point in time, live or simulated live, which engages a unique, targeted audience. Digital events are a relatively new and rapidly changing element in an organization's event marketing mix. According to an MPI report titled "Lesson Learned: The Strategic Value of Virtual Meetings and Events", many companies are beginning to add virtual meetings to their core business processes, particularly for education. According

1947:	Dennis Gabor invents holography in Hungary.
1950's:	Bell Labs introduces audio teleconferencing.
1951:	First Australia School of the Air broadcasts lessons to children by shortwave radio.
1952:	Pennsylvania State University introduces closed captioning television versions of their courses to broadcast programs to remote locations on their campuses to alleviate overcrowded classrooms.
1953:	The Ford Foundation provides a major grant to help develop Educational Television stations (ETVs) in the United States. The first educational television station opens in Houston, Texas.
1957:	First radio satellite communications are launched.
1962	First satellite to relay television communications is produced.
1969:	The Open University distance learning program is founded in Great Britain.
1970s:	AT&T introduces Picturephone video teleconferenceing to multiple sites.
1980s:	Digital telephony is introduced through the development of ISDN lines.
1990s:	Voice over Internet Protocol (VoIP) is introduced to enable computer-based teleconferencing.
1991:	First webcam is developed at Cambridge University.
1992:	First radio webcast. By 1996, there were 86 radio stations broadcasting on the Internet, and in 2013 there are now thousands.
1993:	"Severe Tire Damage" is the first band to perform live on the Internet from Xerox PARC, and scientists webstream the performance to Australia.
2000:	Video-telephony is introduced through programs such as Skype and iChat.
2004:	Go To Meeting is introduced by Citrix to allow a single desktop computer to teleconference with multiple locations.
2008:	Marriott and Hewlett-Packard sign a deal to create HP Halo Telepresence rooms in select Marriott venues.
2008:	CNN introduces hologram of reporter Jessica Yellen in a studio during US presidential election coverage.
2010:	Increased use of mobile smart phone communications for videoconferencing through introduction of Apple's FaceTime application.
2012:	Tupac Shakur hologram is introduced at Coachella Music Festival and later seen by over 13 million people on YouTube.

Figure 7.2. Historic Development of Virtual Meetings and Events (Source: Goldblatt, 2013)

to Goldblatt (2013), the range of digital events can be defined along an experience and complexity continuum—ranging from a relatively simple one-hour webcast or webminar to complex multi-day hybrid events running concurrent with face-to-face meetings, and ultimately to robust digital experiences serving as immersive and interactive year-round online communities.

WHY VIRTUAL EVENTS/MEETINGS?

Today's business world is global and needs a medium to connect people around world. Virtual events can reach the global marketplace. As the Internet has become the universal communication channel, virtual events often delivered via the Internet can provide a strong tool to reach out to global event audiences. Advancement in technology is happening faster and cheaper than ever before. Technology tools allow key players to communicate using audio, video, and even chat tools, which shrink the distance between companies and their employees and clients for the most personal attention.

There are advantages in adopting virtual events and meetings. The top reason is reducing costs when compared to face-to-face meetings. Decreased travel budgets are common in today's business world, and more and more organizations consider virtual events. Global corporations consider virtual meetings for reducing the travel demand on executives, supporting last-minute meetings and sharing information across multiple markets and time zones. The key advantages are listed below:

- Cost saving (travel)

- Reducing time out of office

- Enabling session speakers and attendees to communicate in real time

- Keeping a meeting that would be cancelled due to budget

- Convenience

- Information-sharing across multiple markets

- Global business operations

- Business continuity

- Supporting last-minute meetings

Wang and Lee (2013) listed additional advantages of VMT especially in regard to educational events/meetings:

- *Synchronous interaction:* Videoconferencing has the unique ability to simulate the richest form of human interaction, namely, face to face. Previous studies in the area of human communications show that more than 60% of all communications are derived from nonverbal behaviors, such as using gestures, emotions, body language, etc. Therefore, live instruction via VMT offers opportunities for learners to develop a high level of interaction and allows them to become more engaged during lectures, thereby enhancing the quality of learning.

■ *Ease of use and convenience:* VMT applications are mostly easy to use, especially for today's learners who are comfortable with technology and can learn a new tool quickly. With minimal equipment (e.g., computer, webcam, microphone, Internet connection), a learner can log into a live lecture session and fully participate in the discussion from any location. Some VMT applications are sophisticated and cross-platform compatible, allowing users to access using mobile devices, such as smartphones and tablet computers. Using VMT not only brings convenience to students but also exposes them to additional tools that they may use in their continued education and in their professional careers.

■ *Green technology:* VMT is not just good for attendees; it is also good for the environment. Since video communications are completely digital and transmitted over the Internet, classes that utilize VMT help the reduction of gasoline consumption and emissions and save time by reducing the number of students, instructors, and experts who have to commute to event venue.

TYPES OF VIRTUAL MEETINGS AND EVENTS

There are various formats, or categories, of technology for virtual meetings. Each format is unique but can be sometimes interchangeably used at the same time. Therefore, it is important to thoroughly review each and determine which type works best for a class/educational session. Here are several ways to consider types of technology:

By virtual event happening time of occurrence:

1) Synchronous = real time

Synchronous VMT enables speakers and attendees to communicate in real time, meaning "synchronous" beyond geographical distance.

2) Asynchronous = recorded and flexible in time

Asynchronous virtual event provides flexibility of time as it is recorded materials (video, audio). It emphasizes its convenience as attendees can access the virtual event any time and at any pace that attendees want.

By communication method:

The most commonly accepted formats include 1) audio conference; 2) video conference; 3) telepresence; and 4) 2D/3D immersion. Each VMT format is served with multiple major VMP service providing vendors and offers the following functions:

1) **Audio/Teleconferencing**

Audio/teleconferencing requires a traditional telephone line or the lowest Internet bandwidth. Visual materials are often delivered in electronic files (e.g. PDF or PowerPoint). When organizing the audio conference, there are several key considerations to be made in planning and coordination. First, an audio conference includes a facilitator and an audience. It is important to control audio input when the facilitator speaks as multiple audio inputs can create howling or noisy background noise that may reduce audio quality. Second, it is critical to address which page of a shared document is being discussed by all attendees as there is no video function on audio-only conferencing. Third, if the audio meeting includes new attendees, it is ideal to have participants state their name first so all attendees know who is speaking, and it also can help a note taker of the conference.

Figure 7.3. Polycom Soundstation IP 7000 (Source: www. polycom.com, used with permission)

Over all this is the least expensive and lowest level of technology required for a virtual meeting. Polycom™ (Figure 7.3) and AVAYA™ manufacture high definition (HD) audio/teleconference equipment.

2) **Video Conferencing**

Videoconferencing facilitates a face-to-face meeting environment across borders, clearing the way for efficient communication and collaborative decision-making both within and between organizations. Today, more than ever, it is proving to be an extremely powerful business tool, transforming day-to-day business operations by helping to increase effectiveness, maximize resources and optimize productivity. Cost of video conference systems varies. Some professional systems can be expensive, but simple and

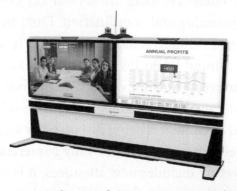

Figure 7.4. Video Conferencing System: Polycom Real Presence Medialign Dual 55 (Source: www.polycom.com, used with permission)

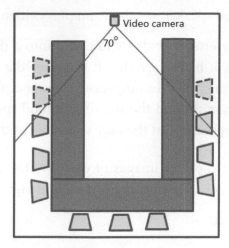

Figure 7.5. Viewable Angle of a Video Conference Room

free solutions (e.g. SKYPE, ooVoo) with limited functionality are available. It is not necessary to purchase high-end systems if you can organize a video conference using cloud-based video conferencing vendors (e.g. CISCO Webex, GoToMeeting).

Advantages of video conferencing include:

- Provides the ability to see rich content (facial expression, body language) of other attendees, which improves effectiveness of communication

- Allows participants to make more frequent contact with colleagues, partners, suppliers, and customers without having to leave the office

- Reduces travel time, stress and expenses

- Allows for ad hoc meetings, which permits the discussion of urgent matters promptly, including immediate decisions

- Allows the sharing visual presentations

Components of a videoconferencing system include:

- Camera and microphone: captures the image and sound of speakers and materials

- ◼ Monitor and speaker: displays and plays video and audio

- ◼ Codec: converts the video and audio into a digital signal and compresses it before sending it out over the network. At the other end, the codec decompresses the signal and feeds the picture to a monitor and the sound to a loud speaker. (This may sound complicated but the user sees none of this.)

As a video camera captures images of each location, logistics (room set-up) must consider a viewing angle as seen in Figures 7.4 and 7.5.

3. Teleprescence

Telepresence is an immersive style of videoconferencing that emulates the experience of having all participants, typically between 5 and 20, in the same room (see Figure 7.6 for an example). The difference between telepresence and traditional videoconferencing is that more senses are involved. The technology transmits crystal-clear audio and high-definition video in real-life size between rooms that are designed to match and appear as an extension of each other. Eye contact, facial expressions, and nonverbal cues are experienced the same way as in real life, making telepresence an immersive, life-like communications experience and engaging, rich-media learning environment.

Components of a Telepresence Room

A telepresence room requires a more advanced and permanent installation, and therefore, it has high infrastructure costs. These include:

- ◼ a telepresence system with built-in touch screen control panel

- ◼ multiple room-size LCD screens that deliver high-definition images with robotic camera clusters that are placed at eye level

- ◼ integrated microphones that provide shielding to block audio interference

- ◼ high-end speakers for reproducing fine acoustic details and thin light reflectors that provide integrated lighting with reduced glare

- ◼ furniture (executive desks and chairs)

Figure 7.6. Telepresence System Equipped Conference Room (Source: www. polycom.com, used with permission).

Companies such as Cisco and BrightCom are leading providers for telepresence solutions in today's market.

3D holographic video conference adopts a 3-dimensional display of the speaker, which is the closest reproduction of real person interaction. Some of the telepresence conferencing are shown here:

https://www.youtube.com/watch?v=3d7sQfIBAwk

https://www.youtube.com/watch?v=jMCR9xep81E

VIRTUAL MEETING TECHNOLOGY (VMT) AND POPULAR SYSTEMS

Innovative Internet applications coupled with improved videoconferencing capabilities have led to the proliferation of VMT. VMT represents the tools for facilitating interactive and synchronous communications that take place over the Internet using features such as audio and video, instant messaging, and application sharing.

Selecting a Virtual Meeting System

Video and telepresence conferencing systems often require a tremendous amount of investment. Therefore, it is even more important to select a reputable vendor.

The following list includes important items that influence the selection of a video and telepresence conferencing system vendor:

■ A sound financial profile and a good business model

■ A strong company history

■ Proven customer service and support and satisfied customers

■ Cost-effective maintenance programs

■ Ethical business practices

■ Enterprise-wide application solutions

■ A true global presence for effective implementation around the world

Table 7.1 summarizes VMT formats, complexity, vendors, and products by expanding the existing contents from MPI Foundation's virtual meeting report (2012) and Lichtman (2006).

Skills/Knowledge Needed to Organize Virtual Events and Meetings

As shown in Table 7.1, there are various levels of complexity in organizing virtual events. However, there are some necessary aspects to consider. They include:

■ Event planning

■ Online marketing

■ HTML coding

■ Digital-user interfacing

Table 7.1. Summary of VMT Formats (source: Adi and Lee, 2013; data from MPI Foundation's virtual meeting report, 2012 and Lichtman, 2006)

Complexity	VMT Format	Description	Vendor	Product
Easy	Web/Video Conferencing	• User friendly-web (video) conferencing technologies • Make the most common types of virtual meetings • Usually limited to 25 or fewer participants • Requires little set-up time	Adobe	Adobe Connect 8
			Cisco	Webex
			Citrix	GoToMeeting
			Microsoft	LLiveMeeting
			SKYPE	SKYPE
			Blackboard	Blackboard Collaborative
Medium	Webcasting/ Webinar	• Include more than 25 participants • Requires more sophisticated functionality and the expertise of a consultant or vendor • Requires days to a few weeks for proper organization • Technological support is necessary	Intercall	Intercall Streaming
			INXPO	XPOCAST
			ON24	Webcasting Platform10
			Sonic Foundry	Mediasite 6
High	Video/ Telepresence	• An immersive style of videoconferencing that emulates the experience of having all participants • Participants, typically between 5 and 20, in the same room • Requires virtual meeting coordinator and an experienced team provided by a technology supplier to deliver this format • Require months to produce	BrightCom	Lumina Telepresence
			Cisco	Telepresence T1/T3
			Polycom	Halo Collaboration Studio
			Teleris	VirtualLive
High	2-D/3-D Immersive Worlds		INXPO	VX Platform
			UBM Studios	UBM Studio
			Unisfair	Virtual Engagement Platform

- Comfort with technology
- Customer service (explaining technology to users)
- Organization (multi-tasking several virtual meetings)
- Flexibility (last-minute needs)
- Thinking quickly, inventively and with passion

In addition, it is very crucial to work closely with in-house or outside IT professionals throughout the virtual event planning and coordination effort. If there is no expert in these areas, hiring a virtual event professional (e.g. digital event specialist) or instructional design specialist is recommended.

NEW TREND: HYBRID MEETINGS

A hybrid meeting involves a mixture of physical events with elements of a virtual event, usually running simultaneously and with overlapping content and interactive elements. Figure 7.7 shows a diagram of a hybrid meeting.

Hybrid meetings have grown rapidly with their combined advantages of both physical and virtual meetings. According to an MPI study

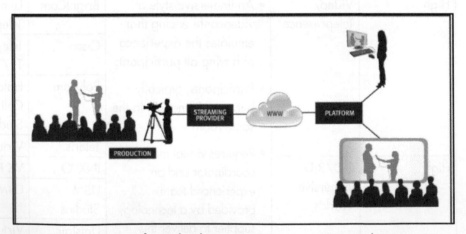

Figure 7.7. Diagram of a Hybrid Meeting (Source: MPI Foundation, 2012; used with permission.).

(2012), this particular meeting format is predicted to grow into an $18.6 billion industry by 2015. Also, it is considered to be an important part of the meeting industry's future as it is considered as a new means of content delivery and engagement.

Advantages of hybrid meetings include:

- broadcasts to remote delegates and remote speakers

- connects remote locations to a main event

- connects multiple sites to a broadcast studio

- creates a legacy after the event

New Professional Designation for Digital Event Planning

As virtual meetings/events are increasingly utilized in the meeting and event industry, the industry seeks skilled event planners who can effectively engage event attendees and others through digital technology. More and more event and meeting professionals express that there is a strong need for professionals who are trained to plan, produce, and measure the results of digitally produced event efforts. To address this need, the meeting and event industry stakeholders have partnered to develop a unique certification. It is titled "Digital Event Strategist (DES)" and designed for professionals who work for various organizations, including associations, corporations, government and academia.

The certification program consists of an application process, an online learning program and an exam administered by VEI. Applicants must fulfill several prerequisites, including:

- current or recent (within 12 months) employment in events, meetings, conventions, marketing, or learning industries

- membership in a professional event management association, such as the Professional Convention Management Association

(PCMA), Meeting Professionals International (MPI), International Association of Exhibitions and Events (IAEE), American Society of Association Executives (ASAE), Corporate Event Marketing Association (CEMA), Virtual Enterprises International (VIE), or three years' experience in event, meeting, convention, exhibition, or learning management

- experience in various virtual and digital environments, such as attending, producing, or exhibiting at online events or meetings

- completion of (or will complete) 25 clock hours (or 2.5 CEUs) worth of digital-related continuing education programs in 16 key areas of focus within 12 months of application

The DES Online Learning Program covers the key areas of focus in the exam through a 16-module course in digital event planning and production. Applicants earn CEU credit for each completed module, for a total of 16 hours in digital event training. Applicants must pass the exam, administered online, within 12 months of application. Because digital technology is constantly evolving and changing, all participants must recertify every two years and fulfill the continuing education requirement of 20 hours over two years. No exam will be required if certification is maintained.

Read more at http://www.pcma.org/connect-and-grow/chapters/greater-midwest/chapter-news/2013/05/29/digital-event-strategist-certification-is-here#.VHQevfldWSo#ixzz3K3gScC2Q

SUMMARY AND CONCLUSION

This chapter contributes toward a better understanding of the emerging VMT technologies that will impact meetings and events in the near future. The role of virtual meetings is evolving although doubters worried that this technology faced too many challenges. In an era of fast broadband, mobility, and flexibility, VMT offers an effective, affordable, green solution to many challenges faced by today's IT educators. More and more organizations are organizing virtual meetings, and others are showing interest in hosting one.

As the industry continues to experiment and gain more experience with virtual events and meetings, this format of virtual events will provide a tool for event professionals to offer new and creative events that add value to their businesses. The various VMT options and formats are utilized by meeting/event planners according to goals, set-ups and the profile of attendees. The successful application of VMT in different event/meeting set-ups and exemplifying the use of VMT as an effective technology can be an enabler of active engagement.

In an era of fast Internet and mobility, VMT can offer an effective and affordable solution to challenges, such as inviting keynote speakers and the latest version of training to a physical session room. Therefore, today's planners should consider and utilize VMT in terms of ways to connect attendees beyond the four walls of meeting rooms as it can open a door to provide effective tools for improved attendees' meeting experiences.

DISCUSSION QUESTIONS

1. What are the top reasons for organizing virtual meetings?

2. What are the key determinants of selecting a virtual event technology system?

3. What is the future of virtual meetings and events and the challenges event meeting planners face?

TASK

Organize a virtual web (video) conference using cloud-based video conferencing providers (e.g. WebEx, GoToMeeting—they both offer free trial for 30 days) with one-end (one-on-one) and multi-point (multiple sites) and discuss the outcome of the virtual meeting, including any challenges.

SOURCES

CISCO (n.d). Video conferencing, accessed 19, March 2014. Viewed: http://www.cisco.com/c/en/us/products/conferencing/index.html

Goldblatt, J. (2013) *Special Events, Creating and Sustaining a New World for Celebrations,* New York, NY:John Wiley & Sons, I nc.

Hrastinski, S. (2008). The potential of synchronous communication to enhance participation in online discussions: A case study of two e-learning courses. *Information & Management.* 499–506.

Kock, N. (2005). Media richness or media naturalness? The evolution of our biological communication apparatus and its influence on our behavior towards e-communication tools. *IEEE Transactions on Professional Communication* 48(2), 117–130.

Adi, Koumae Aubain, & Lee, S. S. (2013). An Exploratory Study of Synchronous Virtual Meeting Technology (VMT) Embedded Classes and Key Factors Influencing Perceived Effectiveness of Various Class Set Ups. Proceedings of the 20th Annual Graduate Student Research Conference in Hospitality and Tourism, Tampa, FL.

Lichtman, H. S. (2006). Telepresence, effective visual collaboration and the future of global business at the speed of light. HPL, Human Productivity Lab Magazine

MPI Foundation (2012) The Strategic Value of Virtual Meetings and Events. Accessed 22, March 2014, Viewed: http://www.mpiweb.org/Portal/VirtualEvents.

Robert, L. P., & Dennis, A. R. (2005). Paradox of richness: A cognitive model of media choice. Professional Communication, IEEE Transactions on, 48(1), 10–21.

Wang, D. Y., & Lee, S.S. (2013). Embedding virtual meeting technology in classrooms: Two case studies. Annual Conference in Information Technology Education (SIGITE), Orlando, FL

ADDITIONAL RESOURCES

CISCO (2010) Tandberg is now part of CISCO, Accessed 1 March, 2014, Viewed: http://www.cisco.com/c/en/us/solutions/telepresence/ttg.html

CHAPTER 8

Evaluation of
Meeting and Event Technology

> *"You can only manage what you measure."*
>
> *The Walt Disney Company*

LEARNING OUTCOMES

As a result of reading this chapter, you will learn how to:

- Measure the return on investment of your meeting and event technology decisions

- Evaluate the return on the marketing investment of your meeting and event technology decisions

- Analyze the user friendliness and usage of your meeting and event technology operations

- Strategize new ways to invest in meeting and event technologies at lower cost and higher return on investment

- Continuously improve the user experience through meeting and event technology

- Anticipate the future mobile and broadband capacity for meeting and event technology users

■   Forecast future demand for meeting and event technology by your staff, attendees, and suppliers.

INTRODUCTION

The Walt Disney Company planned to create one of the most technologically advanced attractions in the world through the development of the Experimental Prototype Community of Tomorrow (EPCOT). Walt Disney envisioned EPCOT as a place where people would live and work with the best technology provided by industry. Although EPCOT was first conceived by Walt Disney in the 1970s along with the planning of Walt Disney World when it finally opened in 1982, EPCOT was a vastly different proposition.

Following Walt Disney's death, the planners at Walt Disney World did not believe they could bring to fulfillment Disney's original dream without his inspiration. Therefore, they redesigned EPCOT to both showcase technology (Future World) along with world cultures (World Showcase) through a series of pavilions sponsored by individual companies.

The iconic centerpiece of the attraction is the geodesic dome entitled Spaceship Earth, which is based on the designs of the legendary U.S. architect Richard Buckminster Fuller. Spaceship Earth has featured for over thirty years early demonstrations of Skype and Google Hangouts as well as other futuristic communications technologies.

Fuller once recalled seeing a giant ship sail by and experiencing an epiphany:

Something hit me very hard once, thinking about what one little man could do. Think of the Queen Mary cruise ship—the whole ship goes by and then comes the rudder. And there's a tiny thing at the edge of the rudder called a trim tab.

It's a miniature rudder. Just moving the little trim tab builds a low pressure that pulls the rudder around. Takes almost no effort at all. So I said that the little individual can be a trim tab.

Society thinks it's going right by you that it's left you altogether. But if you're doing dynamic things mentally, the fact is that you can just put your foot out like that and the whole big ship of state is going to go.

So I said, call me Trim Tab. (Buckminster Fuller Institute, 2014)

Although EPCOT evolved from Walt Disney's early dream of a prototype community of tomorrow, it continues to educate, innovate, and transform public opinion regarding technology through its state-of-the-art information architecture. The "imagineers" at Walt Disney World recognized early on that constant evaluation and intervention (trim tab) was required to ensure the positive evolution of their world-renowned attraction. The same thinking must apply to your approach to meeting and event technological evaluation.

Measure the return on investment of your meeting and event technology decisions

According to Jack Phillips, co-author of *Return on Investment in Meetings and Events: Tools and Techniques to Measure the Success of all Types of Meetings and Events* (Phillips, Breining, and Phillips, 2007), ROI is the financial measure classically defined as the earnings divided by the investment, times 100, expressed as a percent. Essentially, this shows the monetary return on investing in a particular project or program.

A meeting or event is both a project and a program. Therefore, when you measure the ROI of your meeting or event, you are finding out as precisely as possible what your meeting or event is costing and what your organization is receiving in return for this investment.

Technology enables you to measure the ROI of your meeting and event through calculating the exact costs and the expenditures by all of your stakeholders (suppliers, sponsors, as well as attendees) and compare these expenditures to the return you are receiving from hosting this meeting or event.

$$\text{ROI \%} = \frac{(\text{Return} - \text{Cost of investment})}{\text{Cost of investment}} \times 100$$

Figure 8.1. Return on Investment Formula

The basic calculation for ROI is shown in Figure 8.1.

When you invest in new technology for your meeting or event, the rate of return should be higher than the meeting itself. For example, achieving a 200% rate of return over a four-year period is optimal. AXIA Consulting (2014) recommends that you also remember the following additional benefits when measuring the return on investment from new technology, such as

- travel reduction through online meetings replacing face-to-face meetings, with remote support replacing onsite support,

- time saved because of increased productivity and reduction in time to complete tasks

- time saved from reduced length/number of customer service calls

- time saved from reduced numbers of errors

- time saved from improved system reliability and having less maintenance or fewer problems to resolve

- time saved with improved software vendor support and, therefore, quicker responses and faster fixes.

The Association of Corporate Travel Executives (ACTE) reported in 2013 that nearly 90 percent of travel executives stated that mobile technology was having a moderate to significant impact on their travel program (AXIA Consulting, 2014).

The survey conducted by ACTE also confirmed that nearly twenty-five percent of corporate travel departments have a mobile policy for corporate travel. Nearly 40 percent of corporate organizations provide mobile devices for their employees.

However, according to the ACTE respondents, only 24 percent of the corporate organizations surveyed actually encourage their employees to use travel applications. Nearly 32 percent of the companies do not have a policy for the use of travel applications.

The ACTE study also confirmed that using a mobile application to look for and book travel is just one piece of a highly complex corporate travel puzzle. The respondents stated that often they are more comfortable using e-mail or contacting the travel desk of their corporation to conduct the travel booking transaction. However, using a mobile application is preferred for simple or last-minute travel booking.

ACTE officials concluded from this study that as more employees use smartphones to manage their leisure time, the demand for mobile in business travel is higher than ever. By participating in policy development, corporate travel executives can ensure mobile use does not become excessively costly or unwieldy to manage. Governed by a well-developed policy, mobile can then become a hugely valuable asset to their business, keeping their employees informed and giving them unparalleled flexibility when traveling to and within their destination.

In addition to these tangible benefits, there are many other intangible benefits that accrue from the use of technology to measure the effectiveness of meetings and events.

Brynjolfsson and Hitt argued in 2000 that computers are not merely number crunchers; rather they are symbol processors. They are used to organize, transmit, and algorithmically transform any type of information that may be digitized, including numbers, text, video, music, speech, programs, engineering diagrams, and much more.

As computer technology has rapidly and exponentially grown since 2000, so have the intangible impacts and benefits for meeting and event participants. The terms "user friendly" and "user experience" are frequently used in the vernacular of personal and mobile computing technology to signify the overall importance of satisfying the needs, wants, and desires of the end user.

Abraham Maslow is well known for his psychological theory of a hierarchy of human needs. Maslow (1954) places self-actualization at the apex of this hierarchy. However, today self-actualization may also be measured in terms of connectivity, bandwidth, and data download speed. Therefore, when designing and selecting appropriate technology for your meeting and event, it is equally important to remember to measure the human needs and feelings that will result from the incorporation of this technology. In order to measure the human experience as related to technology, you may wish to consider the following key questions:

- First, what are the expectations of your meeting and event participants? Do they expect to experience Wi-Fi everywhere or are they content to use Wi-Fi zones only for stronger connectivity?

- Second, what hardware and software will your participants be using? The types of applications they may use could determine speed of your Wi-Fi signal and could also lead to creating challenges with their connectivity and speed.

- Third and finally, what are the key performance indicators for each participant in terms of their individual return on investment and return on objective?

The term "return on objective" is relatively new in terms of performance measurement and encompasses a wide range of factors. These factors may include but not be limited to networking, education, recognition, and reward. The factors may directly relate to the technological resources available before, during, and after your meeting and event.

Figure 8.2 depicts how at every stage of Maslow's well-regarded and often-cited hierarchy of needs, the technological return on objective may be measured in human terms. Therefore, it is increasingly important that you consider identifying the key factors each of your participants identifies as important prior to finalizing the technological resources for your meeting and event.

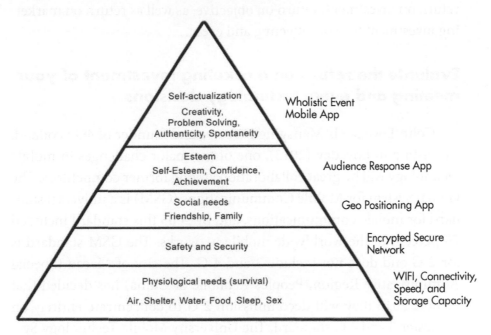

Figure 8.2. Maslow's Hierarchy of Needs (from Maslow, 1954; adapted in Goldblatt, 2014)

One tool for conducting this measurement satisfaction with technology is MeetingMetrics. MeetingMetrics, based in New York City, provides thousands of questions for meeting and event planners to measure return on objective, as well as return on investment. The system they use is similar to those that are available for free (SurveyMonkey and Zoomerang) or at reduced cost; however, MeetingMetrics provides thousands of pre-written questions that allow you to compare your meeting or event to thousands of others that are recorded by this firm.

Furthermore, the MeetingMetrics system includes a pre-survey and a post-survey as well as a pulse phase to allow you to recheck your results the closer you are to the actual meeting or event date.

Regardless of which electronic survey system you use, it is critically important that you comprehensively collect the information you need to make smart technology decisions to effectively measure the return on investment, return on objective, as well as return on marketing investment for your meeting and event.

Evaluate the return on marketing investment of your meeting and event technology decisions

Colin Loveday is Managing Director and founder of 4G Scotland. According to Loveday (2013), one of the major challenges in mobile technology will be great collaboration among former competitors. The Global System for Mobile Communications (GSM) is a universal standard for mobile communications, and by 2005 this standard included 75 percent of the worldwide mobile networks. The GSM standard is for 2 G and does not include 3 and 4 G. The city of Macau (Special Administrative Region, People's Republic of China) has decided that in June 2015 they will decommission 2 G to concentrate entirely on the faster 3 and 4 G standard. The University Mobile Technology Systems (UMTS) is a third general mobile standard that is used widely in the United States and the United Kingdom. Loveday believes that more and more former competitors will work together to achieve a common standard to improve operations and marketing performance for mobile technology.

The use of technology to make us become more human is, according to Loveday, one of the future developments we may expect from meetings and events technology. This may include interaction with 3D printers, robotics, and other emerging technologies before, during, and after the meeting and event. He states that the successful meeting and event marketer must first understand human nature and then develop the technological platforms and solutions to support future needs, wants, and desires.

According to Christopher Preston, author of *Event Marketing: How to Successfully Promote Events, Festivals, Conventions, and Expositions, 2nd Edition* (2012), there are six Ps that impact the effectiveness of your marketing plan. These include:

1. Product: Your meeting, event, exhibition, incentive program, etc.

2. Place: The physical location (destination and venue of your event

3. Price: The price (registration fee, exhibitor fee, other fees) as determined by perceived value and competitive forces

4. Public Relations: The research and information you disseminate to create interest and build credibility about your meeting or event

5. Promotion: The advertising and direct sales activities that will promote sales transactions for your meeting or event

6. Positioning: The unique position your event secures in the market place.

The first five Ps may each be linked to emerging technologies by creating a strong brand (product), a sense of physical space (floor plan), an added value to promote price acceptance (free application use), a third-party endorsement (social media), and electronic transactions (website).

However, the most powerful for the modern meeting and event planner is the sixth P where you may position your meeting and event as the most technologically advanced, friendly, and valuable among all of the similar choices they may have in terms of meeting and event investment. Positioning your meeting or event as a unique value due to your investment in the most effective and valuable technology will best serve your guests and exhibitors.

The measurement of this value is often represented as return on marketing investment (ROMI). ROMI systematically measures the percentage of return on your meeting and event technology investment. For example, if you invest $1000 in marketing your new technology application to your guest and exhibitors, you may also budget to generate new sales (registration fees, exhibitor fees) that provide a five to seven or more return on this original investment. It is the responsibility of the meeting and event planning organization to establish the desired return on marketing investment.

However, there are two types of return on marketing formulas. Short term is a simple index measuring the return of marketing dollars for every one that marketing spends. Long term may be used to determine less tangible benefits of marketing. For example, it could be used incrementally to measure brand awareness over time or to recall the names of key sponsors for your meeting or event.

Meeting and event technology will most often use a long-term approach to measuring return on marketing investment as this will measure the tangible as well as intangible benefits you derive from your technology marketing investment. Table 8.1 provides an example of the long-term intangible and tangible benefits that you may measure using different sets of metrics.

In most cases, due to the rapidly expanding use of technology in meetings and events, you must use the long-term approach to accurately and comprehensively measure your return on marketing investment. You must also be aware of the exogenous variables, such

Table 8.1. Long-Term Tangible and Intangible Benefits of Technology Marketing Investment

Marketing Technology Investment	Tangible Benefit	Intangible Benefit
Web design and management	Transactions	Brand awareness
Exhibitor online registration	Increased square footage	Improved exhibits
Networking mobile application	Improved attendance	Satisfaction
Auxiliary events (tours, etc.)	Increased sales	Faster sales
Virtual education	Increased registrations	Improved learning

as the macro economy may also impact your sales and registration growth year upon year. Therefore, it is often better to measure ROMI over a multiyear period such as three, four or five years to determine an average return from which you will set new realistic targets in the future.

One way to continually increase your return on marketing investment through technology is to plan to provide the most optimum user-friendly experience for your guests and others. This may be challenging due to the wide range of technological experience that your participants bring to your meeting and events as well as the limited technology architecture that is available for your use.

Analyze the user friendliness and usage of your meeting and event technology operations

The term "user friendliness" was probably coined in the late 1970s and early 1980s as geek speak for "keep it simple, stupid." The reality is that no matter how effective or efficient your technological solution may be, someone must first want to engage with it, and when they do they must find the experience simple, easy, and successful.

Is It Simple?

When designing technological solutions for your meeting or event, you may wish to ask yourself if your new application, software, or other solution will pass the "elevator test". Imagine you are on the first floor of a five-story office building. By the time the elevator rises from the ground floor to the fifth floor, can you simply explain the purpose, function, and applicability of your technological solution to others? If you find this difficult, you may need to simplify your technological solution so that your board, your staff, and all of your stakeholders appreciate the value of your decision. Here is one example of how to simplify the description of your new technological solution:

"We are using Cvent's CrowdTorch system so that we have one overall solution for ticketing, mobile communications, engaging with our participants, managing our website, improving networking for our members, and collecting comprehensive data that will help us make better decisions in the future."

Is It Easy?

Some of your stakeholders may understand the purpose and function of your new technology, but they may face a severe obstacle when it comes to applying this technology to their daily needs. Therefore, it is helpful if you provide a short tutorial, webinar, or other instructional opportunity to introduce the new technology to your stakeholders. It is also helpful to identify technology champions who appear to be more confident so that they may mentor others.

Are We Successful?

Success is a relative condition based upon the growing confidence your participants demonstrate through repeated use of your technology. However, to develop this courage within your participants they will need encouragement. Therefore, as soon as you begin using the new technologies, share your success stories to all users through your website, a blog, e-mail, or other communications. It is also helpful to have group face-to-face meetings and have people discuss the benefits of using this technology and to ask any questions regarding challenges they may be experiencing. Finally, you may wish to begin collecting user data as soon as possible and share this quantitative and qualitative feedback with all participants so that they see that your new technology is being embraced and is achieving the goals and objectives you have established. Further, you may wish to express these successful outcomes in terms of return on marketing investment so that your stakeholders recognize the business case for adopting this new technology.

Strategize new ways to invest in meeting and event technologies at lower cost and higher return on investment

Recently, Janet Yellen, the first female chair of the U.S. Federal Reserve, was asked about perceived economic recovery taking place in the United States. She soberly stated that "The recovery still feels like a recession to many Americans" (2014). Although there has been some recovery and stabilizing of financial markets, most meeting and event planners are being asked year after year to continually do more with less.

There are five ways that you may systematically reduce your overall costs when investing in meeting and event technology.

- First, you may wish to use freeware and/or shareware. According to the Free Software Foundation (2015), an organization designed to promote the use of free software and defend user freedoms, freeware is a loosely defined category that has no accepted definition although it is distinguished from free software. Some popular examples of closed-source freeware include Adobe Reader, Free Studio, and Skype.

  Shareware is similar to freeware; however, usually after a prescribed trial period, you must pay for some additional functionality. Still another form of shareware is available only for educational or personal usage.

- Secondly, you can purchase off-the-shelf technology products that may be easily customized for your meeting and event business needs. For example, you may wish to purchase contact management software to manage your registration and then customize this product with data fields that are specific for your organization's needs. If you have a limited budget, you will not want to invest in hiring a software developer to create a customized solution for you from scratch.

- Thirdly, you may wish to partner with other meeting and event organizations to reduce your costs. The International Festivals and Events Association offers an evaluation technological system for festivals that was first developed in Australia. This system is offered at a lower cost because it is jointly shared by many festivals and event organizers throughout the world. Further, if you use this Encore Evaluation System software, you will then be able to make some direct comparisons with the results from other festivals and events.

The cost of new technology generally declines over time. This is caused by the factors of supply and demand. As rival products enter the marketplace, the original products often lose market share and must lower their prices to remain competitive. For example, when the first tablet was introduced by Apple, the cost was well over $700. However, in 2014 similar Android tablets were available at a cost of under $100.00.

Although costs will generally be reduced over time in the marketplace, the user experience must continue to be improved if you are to develop loyal and committed meeting and technology stakeholders.

Continuously improve the user experience through meeting and event technology

To improve any customer experience it is important that you and your meeting and event organization continually look to the future to see what is on the horizon in terms of future stakeholder needs. One way to do this is to monitor the external environment and scan for changes in global business and popular culture.

The rapid movement from the personal computer to the tablet and mobile telephone has been well chronicled. What has been given less attention is how these tools will continue to support the needs of a rapidly aging group of users. According to the United States Department of Health and Human Services (2014), persons over 65 years old numbered 39.6 million in 2009 (the latest year for which data is available). They represented 12.9% of the U.S. population, about one in every eight

Americans. By 2030, there will be about 72.1 million older persons, more than twice their number in 2000. People 65+ represented 12.4% of the population in the year 2000 but are expected to grow to be 19% of the population by 2030.

How will these tools be adapted to support a population whose eyesight may be diminished and who may not have the motor skills for rapid texting as is the norm today? Technology developers are certainly aware of these demographic shifts and are preparing products that will provide additional support for these aging baby boomers. For example, the Calouste Gulbenkian Foundation (2013), UK, in conjunction with *Independent Age*, reported in 2013 that there in some areas, a startling 70 percent of people over the age of 65 had never used the Internet.

Therefore, technology usage is different among demographic groups and is characterized most often in the popular media with two subgroups, called *digital natives* (those in Generation X and the Millennials who were born after the development of key computer technologies) and the *digital immigrants* who were born before the development of key technologies. However, what is often overlooked is that many of the digital immigrants were also the inventors and developers of these technologies.

Regardless of whether your meeting and event participants are digital natives or digital immigrants, one thing is for absolute certain: they will all age and their technology needs will change over time as a result of the aging process. Therefore, it is important to practice the ABCs of growing the user experience over the working lifetime of the meeting and event participant.

- ■ **A** is for Access. It is important that you provide the widest possible meeting and event technology. The same application that works well on your personal computer does not necessarily work as well on the mobile unit. Therefore, it is important to make certain your pre-, during and post-meeting and event technology applications are designed for the widest possible usage.

■ **B is for Broadband.** Connectivity and speed will continue to dominate the needs of meeting and event participants. Therefore, you must plan to satisfy these key needs by conducting a thorough assessment of the numbers of users and the types of applications they will be using, and then provide the broadband capacity needed to support these applications.

■ **C is for Communications.** According to *Independent Age* (2013), the number one desire of older people for technology is to connect with others. This may result from the often-reported sense of isolation and loneliness that results from aging. As we age we often experience physical distance from our children and other family members, and in some cases users may suffer the loss of a partner or spouse. Therefore, the opportunity to connect, communicate and collaborate with others becomes a paramount need in our lives. Your online education, communication, and other relationship-building tools may result in not only improving the user experience but also in expanding the number of participants in your meetings and events in the future.

While these ABCs of growing and improving the user experience are not comprehensive, they do demonstrate how important it is to promote relevance and speed when designing future technologies for meeting and event participants. The number, range, and cost of new meeting and event technologies will most likely expand more rapidly in the next five years than they have in the past fifty years. Your guests will have many different choices. Therefore, to remain a preferred provider of meeting and event experiences for an aging population, you must pick up your pace. One way to do this is to identify the future mobile and broadband needs and capacity for future generations of mobile users.

Anticipate the future mobile and broadband capacity for meeting and event technology users

According to Corbin Ball, in 2014 there were more smartphones in use on planet earth than people. As the number of mobile devices expands exponentially in developing countries such as Africa, it is also important to provide the broadband infrastructure to support these growing needs.

Toby French (2015) observed that at the end of March 2013, there were 654.6 million fixed broadband lines across the world. This represents a growth of 2.0% in the quarter, which was broadly consistent with growth in the previous three quarters. There has been a significant drop in copper cable lines and a rising demand for fiber optic services.

This global broadband growth is not evenly distributed throughout the world. The fastest growth in broadband capacity is in East Asia. However, when these statistics are analyzed by technology within each region, it becomes obvious that fiber optic transmission is rapidly expanding globally, although in the Americas most broadband is delivered via cable.

The top five broadband countries in 2013 were China, the United States, Japan, Germany, and the Russian Federation. China was the dominant country, with 184 million subscribers, followed by the United States, with 94 million. The global growth in broadband subscribers between Q3 2011 and Q1 2013 was 2.5 percent, and in 2013 the fastest broad band growth was in Asia and the East, followed by Europe and then America (Point Topic, 2013).

Of the top ten broadband countries in the world, India was in tenth place, with 15 million subscribers (Point Topic, 2013).

From this dramatic growth in broadband usage in the developing world, North American and European meeting and event professionals must recognize the significant potential future threats to their programs if they do not continue to place pressure upon national and local governments to increase broadband capacity to benefit their delegates. Perhaps one future lobbying issue for the modern meetings and events industry will indeed be technological infrastructure as the meetings and events industry recognizes that they must clearly and convincingly convey to government leaders the need for this expansion.

While in the period from 1970 through 2000 the main effort placed in lobbying by the hospitality industry in North America may have been upon the need for the physical expansion of conference centers, in the twenty-first century the new and even greater effort may need to be placed upon the need to significantly increase broadband capacity for visiting delegates. The competitive edge for destinations may no longer be physical space but instead will be the ability to provide stable, widespread, and fast and unlimited broadband connectivity and storage space for meeting and event technology users. The meeting and event planners' ability to forecast and satisfy these future demands from their participants may determine the winners and losers in the race for dominance as best destinations to conduct meetings and events throughout the twenty-first century.

Forecast future demand for meeting and event technology by your staff, attendees, and suppliers

Historically, future demand has been predicted through historical data and a strategic audit of the external environment in which meetings and events take place. This may no longer be true in the future when major external global forces such as technology, global warming, and economic uncertainty and migration patterns strongly influence the growth of the modern meetings and events industry.

Figure 8.3 depicts how these forces may impact future demand factors in the global meetings and events industry in the next ten years.

TECHNOLOGY

As technology continues to grow and dominate our lives, it is important for meeting and event professionals to anticipate this demand by seeking opportunities to provide differentiated, unique, fast, and easy-to-use as well as stable products and services to improve the lives of their participants.

GLOBAL WARMING

The well-documented rise of global temperatures has resulted in dramatically changing weather patterns, higher energy costs, and the potential relocation of meetings that are associated with changing business and leisure travel patterns. Therefore, meeting and event professionals will need to respond to these changing patterns by providing alternative locations and venues, and wherever possible, by reducing travel distance to demonstrate their commitment to reducing the global carbon footprint. Technology may be used where appropriate to reduce the damaging effects of global travel. However, technology must not be used as a replacement for those critical face-to-face meetings and events that require human contact to achieve the best outcomes. Therefore, in the future, meeting and event professionals will need to carefully balance the need for face-to-face experiences with the damaging impacts of global travel. Meeting and event attendees in the future will be better educated about the damaging impacts of travel and energy costs and

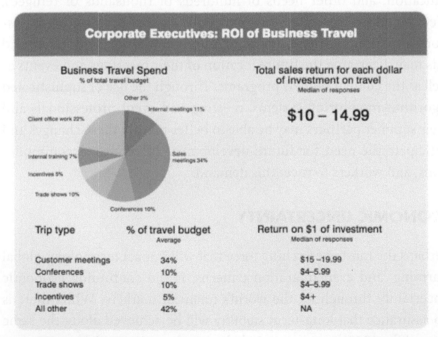

Figure 8.3. Global Forces and Their Impacts on Meetings and Events (Source: Goldblatt, 2014)

therefore will consider these factors when selecting which meetings and events they will support. Therefore, there may be growing opportunities for technology to assess, project, forecast, and determine the negative and positive impacts of global travel to meetings and events. Used properly, these technologies may also be able to increase the demand for more appropriate face-to-face meetings supplemented by technology.

MIGRATION PATTERNS

Warfare, famine, economic deprivation, and other forces have increased the level of global migration to unprecedented levels in the past decade. As a result of these increased migration patterns, neighboring nation states who receive asylum seekers are often hard pressed to provide for their needs in a sustainable manner. This may impact the future of global meetings and events by changing the resources that are available within certain countries (a recent example would be Jordan) as they seek to cope with providing for complex health, education, and other needs of hundreds of thousands of refugees. Meeting and event technology may be leveraged by meeting professionals to forecast where future migration patterns may develop, and this may determine the future location of their meetings and events as well as the content of their programs. Through the use of sophisticated algorithm monitoring systems, meeting and event professionals and their supplier partners may be able to better predict these changes and anticipate the need for future development of venues, accommodations, and workers to meet this demand.

ECONOMIC UNCERTAINTY

Perhaps the most overarching force that will impact technology, global warming, and even migration patterns, is the continuing economic uncertainty throughout the world's financial markets. While there is no assurance that long-term stability will be achieved along the same lines of the 1950s postwar period, there is the opportunity to use meeting and event technology to reduce some costs associated with labor, shipping, equipment, instruction, travel, and other key expenses historically associated with meetings and events. Furthermore, in addition to reducing costs, meeting and event technologies, especially those linked to continuing professional education, may generate even greater financial benefits than the traditional face-to-face experiences of the past. Who knows, perhaps one day virtual meetings and events will provide the greater economic lucre to subsidize and ensure the future of face-to-face meetings. Regardless, the modern meeting and event professional must embrace technology as a tool for lowering costs and driving new future revenues to ensure economic sustainability for their meeting and event organization.

While these are certainly not the only four forces that will most likely determine the future sustainability of global meetings and events, they are important considerations with regard to developing technologies. Future technologies will not only provide greater opportunities to forecast demand for meeting and event services but will, even more importantly, provide the flexibility, rapidity, and creativity to address and solve future global problems that will impact meetings and events.

SUMMARY AND CONCLUSION

The evaluation of meeting and event technology must be an ongoing process. This process may begin with the great unknown about what technological needs, wants, and desires your future meetings and events participants will have to select your meeting and event as a preferred provider for their education, networking, and other professional and personal requirements. However, this great unknown, through the use of appropriate meeting and event technologies, may provide you and your organizations with significant long-term savings as well as increased revenues to support other aspects of your mission and vision. Therefore, meeting and event technology, while not a panacea for all of the ills of meetings and events industry in the twenty-first century, is certainly a key and critical factor to help continually improve your competitive position through providing the very best user experience.

DISCUSSION QUESTIONS

1. What is the best way to improve the overall meetings and events participant user experience?

2. How will changing demographics impact the use of meeting and event technology?

3. What are the four forces for global change in meetings and events and how will they positively impact the growth of your organization in the future?

TASK

Develop a 1000–word plan that addresses how you will evaluate the selection of appropriate meeting and event technology. How you will measure return on your marketing investment and how you will use emerging meeting and event technologies and global forces to forecast future demand for virtual and face to face meetings?

SOURCES

Atlas Travel (2014) Measure Return on Investment for Corporate Travel. Reviewed 8, April 2014. Accessed: https://www.atlastravel.com/meetings-incentives/study.phphttps://www.atlastravel.com/meetings-incentives/study.php

AXIA Consulting, 2014. Measuring ROI in Technology, viewed 16, March 2014, accessed: http://www.axia-consulting.co.uk/html/basic_roi_calculation.html

Ball, Corbin (2014) Tech Talk. Accessed: www.corbinball.com

Brynjolfsson, E., Hitt, L. (2000) Beyond Computation: Information Technology, Organizational Transformation and Business Performance. *Journal of Economic Perspectives*. Volume 14, Number 4, PP. 23-48.

Buckminster Fuller Institute (2014). Trimtab. Viewed 3, March 2014. Accessed: http://bfi.org/trimtab

French, Toby (2015), Broadband tariff trends over time – Q2 2010 to Q4 2014, http://point-topic.com/http://point-topic.com/free-analysis-author/toby-french/page/3/, 2015

Free Software Foundation, 2015. Accessed: http://www.fsf.org/

Goldblatt, J. (2014) International Centre for the Study of Planned Events, Edinburgh, Scotland: Queen Margaret University.

Independent Age. (2013) *Older People, Technology and Community.* Viewed: 13, April 2014. Calouste Gulbenkian Foundation, UK, Accessed: http://www.cisco.com/web/about/ac79/docs/wp/ps/Report.pdf

Loveday, C. (2013) Personal communication.

Maslow, A. (1954) *Motivation and personality.* New York, NY: Harper.

Phillips, Jack J., M. Theresa Breining, and Patricia Pulliam Phillips *Return on Investment in Meetings & Events.* Routledge, 2007

Point Topic Ltd. (2013) *World Broadband Statistics Q1 2013.* Reviewed: 14, April 2014. Accessed: http://point-topic.com/wp-content/uploads/2013/02/Point-Topic-Global-Broadband-Statistics-Q1-2013.pdf

Preston, C. A. 2012. Event Marketing: How to Successfully Promote Events, Festivals, Conventions, and Expositions 2nd Edition. Wiley.

U.S. Department of Health and Human Services, Aging Statistics, viewed: 14, April 2014, accessed: http://www.aoa.gov/Aging_Statistics/

Yellen, Janet (2014) *Financial Post.* Reviewed: 14, April 2014. Accessed: http://business.financialpost.com/2014/03/31/fed-janet-yellen-rates/

Part III

Successful Marketing of Meetings and Events through Technology

CHAPTER 9

Marketing with Wikis, Websites, Blogs, and Podcasts

> "The Internet has become a remarkable fount of economic and social innovation largely because it's been an archetypal level playing field, on which even sites with little or no money behind them—blogs, say, or Wikipedia—can become influential."
>
> James Surowiecki, New Yorker columnist and author of "The Wisdom of Crowds" (Surowieki, 2006)

LEARNING OUTCOMES

As a result of reading this chapter, you will learn how to:

- Be part of the power of Wikipedia

- Plan your website structure for better success

- Create website strategy

- Establish and maintain a company web blog

- Use your company blog for marketing purposes

- Use the advantages of podcast as well as the reasons why those forms of creating content matter for your event

WIKIPEDIA

It is nearly impossible to imagine the world without the Internet. For all kinds of information we search online and expect to find the right information with just one click of a button. It is this situation that forces organizations to create content in different forms and upload it online for consumers who are searching. In the beginning it was easy as the information online was not so huge in quantity. Nowadays one of the biggest challenges is not just to find information, but also to find the right information you need. If you do a search online, it is impossible to miss the enormous bulk of results that lead you to the website of the free encyclopedia Wikipedia. As statistics show, Wikipedia articles are a constituent part of 95% of all Google searches. Even poor quality pages in Wikipedia get millions of hits because they benefit from the popularity of the site (King, 2011).

"Wikipedia is a multilingual, web-based, free-content encyclopedia project supported by the Wikimedia Foundation and based on an openly editable model. The name 'Wikipedia' is a portmanteau of the words wiki (a technology for creating collaborative websites, from the Hawaiian word wiki, meaning 'quick') and encyclopedia. Wikipedia's articles provide links designed to guide the user to related pages with additional information" (Wikipedia).

While this chapter was written, Wikipedia was comprised of 4,508,631 content articles, with 32,828,738 pages in total, 831,041 uploaded files, 21,299,258 registered users, including 1,409 administrators. All those users work for Wikipedia without any payment at all. In addition to this, Wikipedia has versions in 285 languages. Before going into details about how to use Wikipedia for an event, we need to mention some basic facts and rules for using Wikipedia for business purposes. First of all, we must bear in mind that it is not a social media service, in spite of the fact that its content is user-generated. It is an encyclopedia, and that's why its materials should contain reference information. Virtually speaking, everyone can add articles, although not all articles will be published. In most cases other people will edit your article or add more information. You should be prepared that there is no ownership of the information on Wikipedia.

The fundamental principles on the basis of which Wikipedia operates can be summarized into five major "pillars":

1. Wikipedia is an encyclopedia—not a free platform for advertising.

2. Wikipedia is written from a neutral point of view—which means that it does not include editors' personal experiences, interpretations, or opinions.

3. Wikipedia provides free content that anyone can edit, use, modify, and distribute—including you.

4. Editors should treat each other with respect and civility— that is, actually, the only means to pave the way for efficient collarative work.

5. Wikipedia does not have any firm rules—it has policies and guidelines, but they are not carved in stone. (Wikipedia)

As we strive to learn how to create content for our event in this enormous online encyclopedia, we have to review the three core content policies which we have to follow or be well-informed if someone breaks them in connection to our event, speaker or topic:

Neutral point of view: All Wikipedia articles and other encyclopedic contents must be written from a neutral point of view, representing significant views fairly, proportionately, and without any bias.

Verifiability: Material challenged or likely to be challenged, and all quotations, must be attributed to a reliable, published source. In Wikipedia, verifiability means that people reading and editing the encyclopedia can check that information comes from reliable sources.

No original research: Wikipedia does not publish original thought; all material in Wikipedia must be attributable to a reliable, published source. Articles may not contain any new analysis or synthesis of published material that serves to advance a position not clearly advanced by the sources (Wikipedia, 2014).

In most cases people will go to Wikipedia to find basic information about your organization, your speaker, your topic, etc., and in case they don't, they may ask themselves why they should consider your event at all. Of course, it is not possible to create a wide range of articles about one event there. Although each event is usually created by an organization, the articles in Wikipedia should be focus on the organization and not the events themselves. The spokesperson for the Wikimedia Foundation says, "Wikipedia is a complex culture, and sometimes it can feel like the free encyclopedia everyone can edit—except me" (Zetlin, 2012). In the beginning many feel a similar disappointment. Nevertheless, it is not very difficult to understand the rules and principles so that you can start writing for Wikipedia.

What kind of information can we add about events?

Of course, it is not possible to expect that each event will attract people's attention and will result in a separate article in Wikipedia, yet if you look at your event bearing the three core content policies in mind, you will find different opportunities. Here are a few of them:

- **The speakers** (provided you have some) are experts in their field. Check to see if there are already any articles about them and if so, you can add information about their next activities with your meeting.

- **The topic** of your event might be the crossing point. There may be enough information at your disposal that you can summarize in an article.

- **The location**, providing that it is an important location with historic, architectural, or cultural significance, can be an entry point. You can either check for some interesting articles, add information, or create your own material.

- **The technology**, if you use new technology, can be a perfect opportunity to write about. You can add information to a published material, for example, if you are doing 3D mapping on an imposing building or you can look up for an article about the building and add a picture as well as some other interesting information.

There are various other opportunities for contributing to Wikipedia to indirectly promote your event. The Chartered Institute of Public Relations (2014) published recently an online guide for PR professionals about the use of Wikipedia. In fact, as this is an encyclopedia, if people are searching for your information, someone among the huge army of editors will add some materials about your topic. As we have already discussed, it is not a social media but a network of people working together, so if you are interested in improving Wikipedia, either you or some of your team members should become a volunteer. By far the best way is to start with articles about something that you love or are passionate about. Think about some factors which will increase notability. In most cases three factors can be outlined:

- **Biographical articles**: People consult the encyclopedia to find biographical information about companies and people. So be ready to offer it, though it should be informative, not marketing-orientated.

- **Trade publications:** These give credibility and support from the professional field.

- **Authority publications:** These will help if you have been mentioned in the well-known publications, such as *The New York Times*, although other popular media will help too.

As you can see, Wikipedia prefers to publish materials about topics that have already been covered by other media and have become a point of interest to people. In any case, your event contains points of interest for broad audiences. You have to consider carefully which facts related to your event can be described and then write it follow Wikipedia's rules closely for adding information.

However, Wikipedia can also be a source of problems, such as when it contains misleading information connected to your events. If you find some inaccurate information, there are a few things you can do. First, you can fix it yourself! Anyone can edit Wikipedia. Just click the "edit" button on the top righthand corner of the page, make the correction, and press "Save page."

In case you are not willing to correct an error yourself, you can go to the topic "How to report a problem with an article, or find out more information" and follow the guidelines. Bear in mind that the corrections will not happen immediately.

By way of a summary, we should point out that Wikipedia is a free encyclopedia that can be used by your potential guests as a referable source of information. To make your information referable, you need to be sure that there is enough accurate and relevant information concerning your events. If one of the major principles of lean management is to be applied here, we can say: Genchi Genbutsu, which means "go and see" in Japanese. According to this principle, in order to understand a situation fully, one needs to go to the 'real place'—where work is done. That means that to start writing for Wikipedia, edit articles on topics that are related to your professional qualification and contribute to Wikipedia by writing articles that will help other people. Using Wikipedia a as source of information about your event is simple if you dedicate a little of your time.

WEBSITES

Even if you have managed to publish interesting materials in Wikipedia, your event still needs its own website. Nowadays websites are an important tool for potential guests to get in touch with your event. The basic rule for designing a good website is closely linked with your ability to meet the immediate needs of your visitors. According to research, 55% of the small businesses use online marketing for events, whereas 98% use their website to reach customers (Barnatalent, 2011). Why do they do this? In short, online marketing is less expensive than traditional marketing, and although it requires time and creativity, it puts small companies on equal ground with larger ones.

Figure 9.1. Trends in the Marketing Mix (Source: ParadigmNext, https://paradigmnext.com/trends-in-the-marketing-mix; Used with permission.)

Your event must have its own website. You must plan carefully the content of the site from a visitor's point of view. A good starting point is to focus on its usability and not its look. In fact, most of websites are created with one-way communication in mind. For the most part we provide the information our visitors will supposedly find necessary and useful, we publish it, then design the layout, and even add pictures and graphs, hoping that this will be sufficient. However, your content must be planned in such a way so as to take full advantage of all two-way communication opportunities available. This means offering visitors tools to interact with you. In Table 9.1 shows some of the available tools, though this does not show all the tools at your disposal.

According to an article by Tony Haile (2014), CEO of Chartbeat, a stunning 55% of users spend less than 15 seconds actively on a page, which means that it is necessary to create content and a design that will attract an average user who will be willing to spend time on your website. Haile also states that numerous facts that we have taken for granted about the web just aren't true. Haile identifies the three myths that should be taken into consideration when planning a web presence. The goal is to create content not just for its own sake but that will prove useful to your potential guest. It is advisable to be aware of the myths, as listed in Figure 9.2.

Although the mere fact of having shares in social media doesn't mean that people have actually read the content they share, a well-designed website offers easily shared options. In Table 9.2 you can see the most popular sharing buttons available. It will prove useful for you see what kind of sharing options your potential guests used most, so as to offer them the same choice. As you will see for yourself, including share buttons is very easy. Data shows that sharing options creates more traffic to your website than "like" buttons.

However, if the content of your website is static, it is not likely to invite people to come back again and again. Finding ways to attract attention

Table 9.1. Two-Way Communication Tools for Your Website

Tool	Benefits
Comments from blogs	People like to see what other people think about your event. It's like a third-party endorsement
Ratings and references	Can be used as proof for the quality of your event
Links to social media pages	Encourages people to interact with you
Social media sharing	Shows your activity and the support you gain
Contact information	An easy and convenient way for people to contact you via different channels—e-mails, Skype, Twitter, instant messengers, etc.
Contact form	To be used only for special requirements; such forms tendsto become a frequent obstacle to the natural process of communication, due to the fact that people want their information at a moment's notice.

> **Myth 1: We read what we've clicked on.** Most people leave your website within 15 seconds. Articles that were clicked on and which engage visitors tend to be actual news.
>
> **Myth 2: The more we share, the more we read.** Fewer than 100 likes and fewer than 50 tweets can create engagement with the content.
>
> **Myth 3: Native advertising is the savior of publishing.** On the native ad content only 24% of visitors took pains to scroll down the page, compared with 71% for normal content (Native advertising is a form of paid media where the ad experience follows the natural form and function of the user experience in which it is placed. It frequently manifests as either an article or video produced by an advertiser with the specific intent to promote a product.)

Figure 9.2. Haile's Myths about Behavior on the Web (Source: Haile, 2014)

time and again is a must. Although it depends on the particular kind of event, whose variety is great, we can summarize the following:

- You can encourage visitors to share, setting the goal of collecting special prizes.

- You can ask visiters to subscribe to watching special videos before anyone else.

- You can stimulate registration by offering greet and meet opportunities.

- You can ask visiters to subscribe to receive special materials.

- You can promise to share snippets of important information on a daily basis, etc.

The goal here is to plan your content, keeping in mind the fact that people need to have a reason to return, and that reason is, doubtless, to find something they consider important.

Thus, everything should be as easy as possible and secure. Once people give you their e-mail addresses, use them carefully and only for the purpose of sending useful information. As was

Table 9.2. Top 10 Sharing Buttons for Websites

Name of Website	Description
ShareThis	Many button styles are offered, ranging from small to large, which can display social reach numbers — i.e., the number of times a post has been shared.
Socialize	Offers 12 of the most popular buttons including Pinterest and Buffer, a relatively new app that enables readers to schedule sharing times in order to increase reach and visibility.
AddThis	Buttons can be personalized to show users the services that they use most often, as well as integration with over 300 services. Works on almost any website, not just WordPress.
Sharexy	Offers the opportunity to customize not only the look of sharing buttons, but their placement, in over 20 social networks, along with sharing via e-mail.
Digg Digg	Different customization choices for sharing blog posts. Support for e-mail and print buttons.
GetSocial	Bloggers can avail of the option to add a floating social media-sharing box to blog posts. The unique feature allows users to add their own custom buttons.
Slick Social	Users can see summaries of sharing metrics via an administrative panel.
Shareaholic	Shareaholic offers three different button styles and support over 100 sites
Sharebar	Sharebar automatically reorients from vertical to horizontal positions on blog posts.
Social Media Widget	Social Media Widget allows users to input their social media website profile URLs and other subscription options to show an icon on the sidebar.

(Source: Adapted from Chaney, 2012).

already mentioned, your content should be presented in a variety of forms; a short video, however, is more likely to grab attention than a written text. Your website is your place on the web, and using it shrewdly, bearing in mind your visitors' needs, is of paramount importance. Still, many websites are built for the convenience of their owners, not of visitors. Provided that everything you work hard for is done for the sake of His Majesty, the user, even without too many special effects, flashy animations etc., your website will be a good place to spend some time.

BLOGS

Research has shown that if you can hold a visitor's attention for just three minutes, they are much more likely to return than if you only hold them for one minute. Though visitors come to get new content, websites, by default, are not created for frequent updates. Many of the traditional sections like—for example, about us, contact us, what we do—are not subject to daily update. To get around that, many companies create company blogs as a place for sharing regular updates and insights. Corey Eridon (2014) from HubSpot says about business blogs: "Business blogging is a marketing channel that helps support business growth. It does that by driving traffic to your website and providing opportunities for that traffic to convert in some way".

At the heart of business, blogging is the unique opportunity to provide a place for conversation between the representatives of the com-

Table 9.3. Blogging Terminology

Blog (noun): a journal or diary that is on the Internet
Blogger (noun): a person who keeps a blog
Blog (verb): to write a blog
Blogging (verb): the action of writing a blog
Blog category (noun): categories allowed for a broad grouping of post topics
Labels (noun): a way to categorize your posts easily. When you're writing a post, click Labels on the side and enter the labels you like, separating them with commas.
Tags (noun): similar to categories, but they are generally used to describe your post in more detail

pany and the potential or real customers (between event organizers and potential guests). The smart choice is to use a URL in the following format: blog.eventcompanyname.com/_____ (specific blog title). By doing so, each single blog post becomes a new page of your website and search engines will welcome that regular activity.

The decision to establish a blog has to be taken carefully. In the first place, a blog needs regular updates—a new post on a daily or biweekly basis. Most professionals agree that companies should publish daily on their blogs, yet most do not manage to. Experience proves that publishing at least twice per week can be effective.

As with all business activities, your event blog needs to establish the right goals. Your next step is to discover what kind of information your visitors want that is not currently available on your website. Then you have to find the right person or persons from your event team who enjoy writing and who will agree to prepare the blog content. Together you have to create a list of potential topics as well as some directions for where to look for topics. In the beginning you need to fix a schedule for your blog posts. Visitors like to have some sense of security; knowing that they can find a new post every Monday and Thursday, for example, will definitely make them feel better. Thus, start with goals that are not so ambitious, but focus on the number of posts per week. It is better to increase the number of posts while the opposite would be bad for your reputation. To grab people's attention, you have to learn how to write catchy headlines as well as present great graphic illustrations, making this part of your regular practice. A very important step is to create a way to measure your success and establish exactly what will be accepted as success: number of visitors, number of comments, numbers of shares, numbers of likes, etc. Unless you establish a measurement system before your start, it will be difficult to know how well your blog is going.

If you are not sure that blogging is for you, consider Figure 9.3.

It will be quite reasonable of you to consider carefully everything that will happen before, during, and after your event as an opportu-

Figure 9.3. Effective Blog Strategy

nity to create content for a blog and spread the word about your event through the blogging platform. What is more, as data show, daily posts are the best choices, yet our advice is you to start with two posts per week, allowing you time to adapt to the new channel and to find topics for blogging, as well as to encourage other members of your event team to start offering topics and even writing posts.

One of the essential differences between your website and your blog is the writing style. In the beginning you will doubtlessly have to write and rewrite your posts. Keep in mind that you can write in an informal style just as if you are talking to your visitors. Actually, your blog is a place for conversations, not for monologues. So a good post should inspire readers to share their thoughts with you and for you to respond. That's how the conversation begins. Take time to consider precisely your blog categories.

What does writing a good blog post involve?

Here are some guidelines to help your blog become an efficient tool for the marketing of your event.

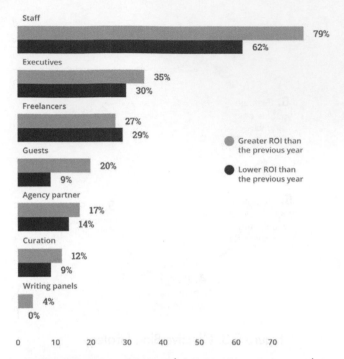

Figure 9.4. Content Creation Sources by ROI. Companies with Lower ROI Call on Fewer Guest Bloggers and Staff Writers. (Source: State of Inbound 2015, Hubspot, stateofinbound.com. Used with permission.)

1. **The title, stupid.** Because your title is the first thing people will see, it should be attractive, provocative, attention-grabbing, and descriptive. At the same time your title must contain your keywords for SEO purposes. It is a well-known fact that the shorter the title, the better its impact on the reader. Fifty-five characters is a reasonable title length limit in accordance with the new design of Google. But consider very carefully where to use capital letters. And if you can fit your title in 50 characters, it will be even better.

2. **Fast reading, the need for short text.** We live in a world in which everything is very fast. People don't have time to read long articles, so divide a long text into many posts, each one no longer than 600 words.

3. **The picture** is worth more than a thousand words. Dedicate time to finding or making a cool picture. Carefully name the title of the picture. You can name the picture after the title of the post.

4. **Format** makes your text easy to read. Include different colors, fonts, quotes, tables, etc.

5. **Categories, labels, tags.** These are there to help your visitors find what they want in your blog, and they also help search engines to "read" your post. Make sure you choose these accurately bearing your SEO in mind.

6. **Linking is the key.** Use your post to guide readers to previous posts or to the content on your website by inserting hyperlinks.

7. **Description**. Use the description as a subtitle whose main goal is to make your post attractive to readers, and, of course, repeat your keywords.

8. **A call for action**. End your post with something that will make people eager to share their thoughts with you. It could be a question, a provocation, or an encouragement to share personal insight.

9. **No one is perfect, even you.** It is a well-known fact that when we write passionately, we often make mistakes. Allow yourself time to read your post and edit. Don't forget that editing is absolutely necessary as well as using a spellcheck program.

10. **Share with the world!** You can use systems to share your blog post on different social media platforms, but make it a habit to make sure your posts are correct.

There are many ways that you can automatically share your content on the Web, and bufferapp.com is just one of them.

As far as blogs and events are concerned, one should think about applying different strategies to cover the event in your blog, as well as to get some bloggers who attend your event involved in various activities.

In most cases people want to share where they are and what is happening. If you prepare a few posts and schedule them to be published every day of your event, adding only new details and photos, this will provide your guests with an opportunity to check in regularly and to share with their friends and followers in their networks. One of the main advantages of blogs is the opportunity to create Internet links, and those links make word-of-mouth contact much faster and more effective in comparison with traditional forms of publicity. It is, in fact, when people trust bloggers that they are inspired to read and share interesting information they come across in blogs. Covering events in this way helps you to create steady traffic to your site and gain credibility. Many of your guests, or at least a few, will write about your event in their own blogs. Take care to offer them support, such as photo opportunities or access to professional photos, and provide them with opportunities for conducting interviews with your keynote speaker or event manager. The latter will make their posts better, which, eventually, results in effective coverage of all your activities and events.

Furthermore, one shouldn't forget that 61% of U.S. online consumers have made a purchase based on recommendations from a blog, and 81% of U.S. online consumers trust information and advice from blogs (Liubarets, 2013). Your blog is the voice of your event or event company in the online world. It will create a community of people interested in all of your events, rather than only one event.

PODCASTS

Having discussed blogs in detail, it's time to focus our attention on one more channel—the podcast. At the begining of the 21st century, the term "podcast" was coined from the words iPod and broadcast. As its name suggests, its content can be used on mobile devices. In general, a podcast is a digital audio (or video) file made available on the Internet for downloading to a computer or portable media player.

Table 9.4. Podcast Terminologies

> **Podcast (noun):** The content in a single episode. The episode contains a digital audio (video) file.
>
> **Podcast (verb):** The action of sending the episode's content out through syndication.
>
> **Podcatcher:** A software application that automatically checks podcast feeds and automatically downloads new items.
>
> **Punchcast:** A podcast that is sent directly to a smartphone or another mobile device, without being sent to a laptop or desktop PC.
>
> **Vodcast:** A video podcast.
>
> **Mobcast:** An audio program that can be received on cell phones or mobile devices.
>
> **Juice:** A free program that can automatically download new episodes/ shows when they become available and can synchronize them with portable digital audio players.

Here are some statistics about podcast users, which are important to consider:

- The percentage rate of Americans who have ever listened to an audio podcast amounts to 29% overall.

- Similarly, 26% of Americans aged 12+ have viewed a video podcast.

- The podcast audience is significantly more likely to have viewed television programming through nontraditional means.

- One in four podcast consumers plug their MP3 players or smartphones into their car audio system "nearly every day." (Webster, 2012)

Those who listen to podcasts are not as numerous as those who log in on their Facebook profile, yet podcast listeners are more focused and will dedicate more time and attention. They will listen to your content for at least 10 minutes. Most podcasts last from 10 to 60 minutes. Imagine what it takes to attract attention for one hour with other media.

Unlike blogs, podcasting is not for everyone because it requires more investment in time, technical skills, and content creating. In the end it does not attract a huge audience, big traffic, or new business.

Yet, if you have interesting content to share and a small but enthusiastic group of people willing to listen, it will be worth all efforts. To start, what you need most is dedication and some equipment.

Equipment includes:

- Audio software. There are different choices according to your preferences
- Microphone
- Mixer and audio interfaces
- Portable recorder/mobile voice recording

However, you can start with just a microphone, a computer, and some audio software; it all depends on your intentions. Besides the

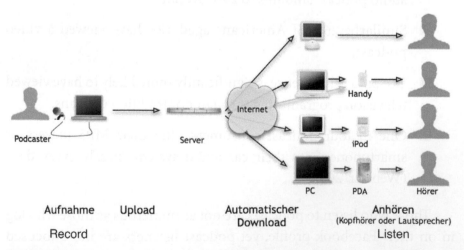

Figure 9.5. How to Produce a Podcast (Source: http://www.content-crew.de/content-marketing/podcasts; used with permission.)

iTunes	To submit your RSS go to the Submit Podcast link.
Podcasts on Windows Phone	Just send an email to podcasts@microsoft.com with the name of the podcast and its RSS feed URL.
BlackBerry Podcast	Create an account with RIM Podcast
Miro Guide	Create an account with Miro Guide
DoubleTwist	Email support@doubletwist.com with your podcast information
Stitcher	Sign up as a Stitcher Partner.

Figure 9.6. Listing of Podcast Directory Sites

technical requirements, you need to have good reasons to start your event podcast. And, last but not least, be patient when you are trying to attract listeners. You have to apply the same strategy for your podcast as for blogging.

First of all, try to learn some of the typical personality traits of your event guests in order to create audio (or video) content that will grab their attention. In most cases interviews with people build better relationships within the community. Moreover, you can tape an interview via Skype or Google Hangouts. Remember that a lot of people who don't have time to read blogs are willing to spare some time listening while commuting, which they are bound to do provided that you offer them exciting content. As for events, there is always something to be shared before and after the event is over. The podcast, when you get used to that format of content, is an easy way to generate guest content. People prefer talking to writing, so even the busiest person will be able to spare ten minutes for a talk with you. If you are well prepared for the interview, you will not need to do much editing. For editing, you can always turn to free options, such as Audacity (http://audacity. sourceforge.net/), for example. When you have an MP3 file, you need just a few minutes to post on your blog, site, Facebook page, G+ page, or wherever you want. Last but not least, the audio content can easily be transformed into written content with the application of transcription software, such as Express Scribe (http://www.nch.com.au/scribe/).

However, do not forget that you have to proof your transcript before posting it on your site.

Podcasts attract influential listeners. What is more, the human voice can have a much greater impact on users in comparison with the printed word. As with all tools in marketing, a podcast has some cons as well:

- It could be costly to create high-quality audio content.

- At the moment all known tools for tracking ROI can't be applied to podcasts.

- If you do not create content on a regular basis, you will lose your listeners

- You may need a dedicated person from your team for production.

- To a certain extent preparing content for a podcast requires creativity and equipment; as a result, not every person from your team will be keen on working on podcast projects.

If creating audio content is not easy for you, you will find video content even more challenging. Like the podcast, a video podcast (or vodcast) is a regular video installment to broadcast messages about your event to the world. The difference is that usually vodcasts are shorter, as they are planned to be watched and, generally, people cannot spare much time. Optimally, video podcasts should be about two to nine minutes long. In addition to this, vodcasts can help you cultivate a base of potential guests for your event. All in all, people enjoy watching short videos. In a blog post titled "25 Amazing Video Marketing Statistics," we can see the following useful information (Mincher, 2014):

- Videos increase people's understanding of your product or service by 74%

- A third of all online activity is spent watching video

- 50% of users watch business-related videos on youtube once a week

■ 80% of your online visitors will watch a video, while only 20 percent will actually read content in its entirety

The numbers mentioned above prove the importance of video content. However, it is very important to make a distinction between a vodcast and video content. Thus, if you decide to produce a vodcast, you should produce do it on a regular basis, as otherwise you will lose your audience.

SUMMARY AND CONCLUSION

Wikis, websites, blogs, podcasts, and vodcasts are just a few of the numerous opportunities you can take advantage of to generate content and start a conversation with your guests. It goes without saying that we cannot be everywhere we want. This means that we have to organize our event by utilizing the new technologies available, which will help not only to make our work easier and faster but will help to establish new expectations from us as event organizers. Online users are not patient enough to wait for information. They want it at a moment's notice and in the format that they prefer. We need to find out who our potential guests are, where they spend their time online, what kind of information they need, and in what format they prefer to get that information.

DISCUSSION QUESTIONS

1. What is your attitude about Wikipedia articles?

2. What kind of information do people look for at event websites?

3. What topics do you find convenient to post on an event company blog?

4. What is the attitude of your peers about audio content?

TASK

Produce an audio interview for a podcast of an event company. How will you go about the interview? How can you convert the audio content into a blog post? What is your idea of a suitable title? Think about how this audio content can be transformed into a vodcast. What are the differences?

REFERENCES

Barnatalent (2011). "Small Business Are More Comfortable With Social Media," December 6, 2011. https://barnatalent.wordpress.com/2011/12/06/small-business-are-more-comfortable-with-social-media/

Chaney, Paul (2012). "Top 10 Social Sharing Buttons For Your Website." Practicalecommerce, November 21, 2012. Http://Www.Practicalecommerce.Com/Articles/3817-Top-10-Social-Sharing-Buttons-For-Your-Website.

Chartered Institute of Public Relations (2014). "Wikipedia Practice Guidance for Public Relations Professionals, Version 2.1". http://www.cipr.co.uk/sites/default/files/CIPR_Wikipedia_Best_Practice_ Guidance_v2.1.pdf accessed 09/10/2015

Eridon, Corey (2013). "What is Business Blogging?," http://blog.hubspot.com/marketing/what-is-business-blogging-faqs-ht, accessed 10.04.2014

Haile, Tony (2014). "What You Think You Know About the Web Is Wrong," TIME, March 9, 2014. http://time.com/12933/what-you-think-you-know-about-the-web-is-wrong/, accessed on 10.04.2014

King, David (2011). "Why Wikipedia is More Important than Twitter." Socialfresh, Nov 16, 2011. http://www.socialfresh.com/why-wikipedia-is-more-important-than-twitter-2/

Lieberman, Martin (2011). "More Small Businesses Willing to Use Social Media, Survey Finds." 11-15-2011 09:30 AM https://community.constantcontact.com/t5/Constant-Commentary/More-Small-Businesses-Willing-to-Use-Social-Media-Survey-Finds/ba-p/41709

Liubarets, Tatiana (2013). "Top Blogging Statistics: 45 Reasons to Blog." Writtent, Apr 25th, 2013. http://writtent.com/blog/top-blogging-statistics-45-reasons-to-blog/

ParadigmNEXT (June 08, 2015). "Trends In The Marketing Mix." http://info.paradigmnext.com/blog/trends-in-the-marketing-mix/

The Podcast Consumer 2012, http://www.edisonresearch.com/home/archives/ 2012/05/the-podcast-consumer-2012.php, accessed on 10.04.2014.

Surowiecki, James (2006). "Net Losses." The New Yorker, March 20, 2006. http:// www.newyorker.com/magazine/2006/03/20/net-losses

Webster, Tom (2012). "The podcast consumer 2012," May 29, 2012, Tom Webster. in Podcast Research, Podcasts, http://www.edisonresearch.com/ the-podcast-consumer-2012/ accessed June 20, 2013

Wikipedia. http://en.wikipedia.org/wiki/Wikipedia:About, accessed on 10.04.2014

Wikipedia. http://en.wikipedia.org/wiki/Wikipedia:Five_pillars, accessed on 10.04.2014

Zetlin, Minda (2010). "Use Wikipedia as a Marketing Tool." Inc. JAN 18, 2010. http:// www.inc.com/managing/articles/201001/wikipedia.html

ADDITIONAL RESOURCES

Ayers, Phoebe, Charles Matthews, and Ben Yates (2008). "How Wikipedia Works: And How You Can Be a Part of It", No Starch Press.

Colligan, Paul, MikeKoenigs, and Gene Naftulyev (2013). Podcast Strategies: How To Podcast - 21 Questions Answered CreateSpace Independent Publishing Platform.

Collins, Lancaster (2014). "Blog Marketing: A Proven Blog Marketing Strategy to Boost Your Business Significantly," Pogo Book Publishing.

Handley, Ann, C.C. Chapman (2012). "Content Rules: How to Create Killer Blogs, Podcasts, Videos, Ebooks, Webinars (and More) That Engage Customers and Ignite Your Business", Wiley.

Nahai, Nathalie (2013). "Webs of Influence: The Psychology of Online Persuasion", FT Press, 2013.

CHAPTER 10

Meeting and Event Social Media and Network Technology Solutions

> *"In the first place social media allows us to behave in ways that we are hardwired for—as humans. We can get frank recommendations from other humans rather than some faceless companies."*
>
> —Francois Gossieaux, Co-author of The Hyper-Social Organization: Eclipse Your Competition by Leveraging Social Media (Gossieaux and Moran, 2010)

LEARNING OUTCOMES

As a result of reading this chapter, you will learn how to:

- Understand and appreciate the importance of social media and social networks for planned events

- Take the best from the differences between social media and social networks

- Decide which social media and social networks to use for the needs of your planned events

- Create and manage content (written, audio, video) for your planned events in the social media and social networks

- Stimulate conversation and engagement between your guests before, during, and after your planned events

- How to measure and evaluate the success of your planned events in social media and social networks.

INTRODUCTION

It was the royal wedding of Prince William and Catherine Middleton in London on the 29th of April 2011 that showed to the world for the first time how new channels of communication have changed the field of special events. The whole world was watching via the official YouTube channel online, and people read and wrote comments, sent video messages to the couple, and did plenty of others things that made the world part of the event.

Nowadays it's very difficult to imagine how people lived in the past without these new ways of communication. Today no one has the patience to wait for the postman to bring the news. We want all information exactly at the moment—the news, the pictures, and the video. Social media and social networks have become the essential platform for social interactions and a permanent means of communication between people. Before going in depth on how to use social media and social networks for the success of your planned event, it is important to look back at its historical development.

Of course everything began with the invention of the technology to send e-mails. As you will see in Figure 10.1, the time frame between the introduction of new social networking technology is getting shorter and shorter. In just four years (2002–2006) four social networks appeared on the landscape, changing not only the way people spend their free time, but also how they communicate or decide what to buy, which events to attend, and in general, how they conduct themselves

Figure 10.1. The History of Social Networking.

as citizens. It is indeed a very short period for such significant changes. The good news is that in this situation we are all equally unprepared. There is no time for research on what is efficient and why, and why people react or do not react to specific messages, images, and incentives. While this text is being written, Facebook has, as of June 2015, 968 million daily active users on average, 844 million mobile daily active users on average, 1.49 billion monthly active users, and 1.31 billion mobile monthly active users (Facebook, 2015).

The average friend count is 190 (Backstrom, 2013), and Facebook is a terrific absorber of audiences' time and attention, with 114 billion minutes a month in the U.S. alone spent on that network. It is difficult to imagine how much time each one of us dedicates to social media and networks. Yet we know that we are there, and we communicate with friends. That is one of the reasons for many event planners to do their best to use social media and social networks in their work: our potential guests are there, and if we want to communicate effectively with them, we have to be there too.

SOCIAL MEDIA AND SOCIAL NETWORKS

We get used to using social media and social networks as two terms, which seem to be interchangeable. The truth is that they are different, and realizing the difference will help us to use them better for the needs of our planned event. According to Social Media Today (Hartshorne, 2010) the differences between social media and social networks are numerous and as vast as the contrast between day and night. We can group all significant differences into five major categories:

What we do when we use both:

Social media: share information with a large audience

Social networks: it's all about creating relationships

Communication model

Social media is a two-way asymmetric model, which means that there is a source of information and people who react to that information liking, commenting, sharing, etc. However, the source remains the same and does not change during to process of communication.

Social networks are a two-way symmetric model as well, with conversations taking place depending on the topic and various interests. In fact, people become engaged. At any time anyone can initiate a new conversation.

We can find one of the simplest explanations of the differences between social media and social networks in the blog by Lon S. Cohen (2009), which explains that "social media can be called a strategy and an outlet for broadcasting, while social networking is a tool and a utility for connecting with others."

Social media and social networks are welcomed by the events planner

Findings from the "Social Media & Events Report 2011" (Amiando, 2011) imply that Facebook and Twitter are the most important social media platforms for event organizers. There are different strategies for using social media and social networks in planned events, yet if we go deep into the strategic management of planned events to see the details, we will discover that in most cases new ways of communication are used for nearly the same purposes. These purposes are as follows:

- **Creating anticipation about the event** is, in fact, a perfect way to share information with potential guests at an early stage of the planned process. If used properly, social networks can help you spread the news at very low cost.

- **Enhancing mutual contacts between guests before the event.** One of the main reasons for people to attend special events is to get in touch with new people, to share ideas with similarly minded people, in other words, become part of a community. With social networks that is easy, and if you succeed in creating a community, you will see that your event will last much longer in the online world.

- **Receiving information from guests in advance.** An event planner sets up the event process by collecting information about the needs and expectations of potential guests. Via social media and networks we can ask questions, collect opinions, and start the conversation with our guests long before the event itself, which will provide a unique opportunity to create an event that best matches the needs and expectations of our guests.

- **Generating excitement and recreation.** Our potential guests can be encouraged to participate in quizzes, award programs, and such, and there are numerous different ways to take part in the preparation process and participate in spreading the news about our event. It's not necessary to give monetary awards. You can award top places, meet with keynote speakers, offer special gifts, etc. What is essential is to use the guest's enthusiasm to generate excitement and expectation.

- ■ **Preparing guests for the event.** Sharing the right information with potential guests in advance—video with speakers, words of welcome, data and any kind of useful information, allows you to prepare guests for what to expect during the event. In that way, we will not fall into the trap of wrong expectations, which is a serious mistake for every event planner.

- ■ **Prolonging the life of your event.** Thanks to social media and social networks we can maintain the community created by the event until our next event. Those people who are satisfied with our last event and actively participate in online discussions are the best ambassadors for our next event. Keeping the community alive is a challenge that event planners have been confronting for a long time, but today there are easy ways to accomplish it.

The basic truth about social networks is that, as they are a form of engagement, you need to be part of the network before spreading the news about your event. This means that each event planner has to take part in online conversations if they want to use that channel for the needs of the events they are planning. You have to choose which networks are most suitable for the kind of events you work on and invest time to create an effective network of people who will be your potential guests one day. Let's have a closer look at some of the most popular sites on the web. Among the most important decisions are which networks to use for the particular event and how to involve guests in taking part in the process of generating content.

PRACTICAL USAGE OF SOCIAL MEDIA AND SOCIAL NETWORKS

FACEBOOK

Of course, we will start with the most popular social network at the moment of the writing of the book. According to eBiz Facebook has 900,000,000 estimated unique monthly visitors (eBizMBA, 2015). That's a huge number, and among all those 900,000,000 the majority of our guests are to be found. As Facebook is changing very often, we will not go into detail over the steps in creating events on Facebook. The truth is that with the new ways of communication, more and more people find new friends as well as new people sharing similar views as their own, and they want to gather and organize an event. There are currently more than 16 million events created in Facebook each month. The majority of those events are offline events, from class-mate meetings to political protests. When you decide which social network (or networks) to choose for your event, it is a good idea to search where your potential guests are. Although Facebook is the most popular network, sometimes the potential guests of our event can be found in quite a different place, such as a less popular network but one that attracts people with specific interests. One example is www.networkingforprofessionals.com, a business network that combines online business networking and real life events.The variety is great, but you have to think choose carefully, keeping the purpose of your event in mind.

Since Facebook changes its features frequently, it is helpful to visit the page on Facebook containing "How to" advice for the latest updates about Facebook. However, whatever the new features are, you have to keep in mind some basic rules:

■ **Start early**. The right way is first to set up your personal network, then create a page for your company and get people to like it, and only after that should you get on with creating events on behalf of the company. The sooner you start, the better chance you will have to build a strong network and spread the news about the events to the right people.

Provided that your event is significant enough or will be repeated in time, it will be of vital importance to create a page for the event with its name and create events on behalf of the page. This will help you to create the community around your event and maintain relationships with those who like your page.

■ **Fill in all the details about the event.** Give users as much information as possible during the early stages and add details as they become available. Don't forget to add an event photo. (For detailed information on how to add photos, visit the page).

■ **Decide on some very important details.** Will all guests be able to see who is invited and who plans to attend and who chooses not to attend? Think carefully, because people may not wish to attend an event to which more than 1000 people have already refused to come. One of the most popular approaches is to show only the guests who will attend, though you have to take a decision depending on the event.

■ **Upload photographs.** People are attracted by pictures, so don't miss that great opportunity. Make sure that all uploaded photographs have arranged copyrights. You can upload pictures connected with the upcoming event such as keynote speakers, artists, place, city, billboards, posters, etc. It is best to upload your photographs immediately after the event; otherwise somebody else may do it. Arrange the photos into albums so they are easy to find and share. Write carefully the name of the event and the date and add a description. It is a good idea to let people tag themselves to some of the pictures, and note not all people will want their names on their photographs. Give them the choice.

Top 10 Most Popular Social Networking Sites | October 2015

	Estimated Unique Monthly Visitors
1. **Facebook**	900,000,000
2. **Twitter**	310,000,000
3. **LinkedIn**	255,000,000
4. **Pinterest**	250,000,000
5. **Google Plus+**	120,000,000
6. **Tumblr**	110,000,000
7. **Instagram**	100,000,000
8. **VK**	80,000,000
9. **Flickr**	65,000,000
10. **Vine**	42,000,000

Data from: eBizMBA Rank , http://www.ebizmba.com/articles/social-networking-websites, accessed 19/10/2015

Figure 10.2. Top 10 Most Popular Social Networks . (Data from eBizMBA Rank, http://www.ebizmba.com/articles/social-networking-websites, accessed 10/19/2015.)

■ **Upload video.** Video information is a must in today's communication process. People are addicted to motion pictures, and they really expect to watch video materials before and after the event. You can tape some welcome words by the organizers, a few words by participants, a video about the place and many other topics. Don't be afraid to experiment. Post a few statutes before starting the invitation process.

Other suggestions:

■ Post links to your official website or blog. Share links with information related to your events.

■ When the page of your event contains all kinds of information, you are ready to send an invitation.

- Encourage your guests to invite their friends.

- Answer and comment on the questions posted on the event wall. That is very important.

- And don't forget that paid advertising is an option here too. You can promote your event with targeted advertising on Facebook.

At the time this book is being written, from an event planner's point of view, there are two main downsides of the Facebook event application. In the first place, the applications do not offer an opportunity to create an event that can be repeated each month, for example. If you have that kind of event, you have to create a new event each time and repeat all steps mentioned above, which is easy yet time-consuming. Secondly, the application doesn't support ticket sales. When you create an event from your business's Facebook Page, you'll have the option to add a link to a website where people can buy tickets for your event.

By far Facebook has been the most popular social network in the last few years. It is still very crowded, with more than 16 million events created on Facebook each month (https://newsroom. fb.com/Products/) competing for the user's attention every month. If you want to manage your event successfully, you have to encourage friends to spread the news. That's what the power of social networks is all about.

LinkedIn

While this text is being written, LinkedIn is the biggest professional network in which people connect based on professional goals. It is the place where people look for professional meetings, seminars, conventions, training, etc. "Most people use LinkedIn to 'get to someone' in order to make a sale, form a partnership, or get a job . . . However, it is a tool that is under-utilized" (Kawasaki, 2011). LinkedIn has currently 259 million active users monthly as of October 2013.

According to Jeff Weiner, CEO of LinkedIn, they "continue to deliver value to professionals through investment in core products and strategic initiatives such as mobile, students and the professional publishing platform."

Thus, if the main purpose of your event is to meet a professional body, LinkedIn is the right place to find your potential guests, though it no longer supports the LinkedIn events application. However, there are still many ways to spread the word about your upcoming events. You can always share links about events from your home page or post a discussion in relevant groups. The first thing to do is to create your personal profile and establish your network. A reasonable step after creating your account is to search for some professional groups in LinkedIn related to the topic of your event. Since this is a professional network website, the level of communication noise is much lower than in Facebook or Twitter, for example. Here people do not receive tons of information about events, so your message stands a much better chance to be noticed. You have to create a home page for your organization and fill it with information before you start to invite people. Think carefully how many discussions can be handled in different groups, as people will be raising professional questions.

Due to the professional nature of the users, it is necessary to send targeted guests a customized message inviting them to your event. As in all computer-based communications, standard messages are not the best way to invite people to your unique event. Take some time to think about what to write and make sure you tick the box to hide recipients' names on the message in case you send one message to a number of people. It is not a good idea to show whom you are inviting or show details of people who may not want their profile to be available to others.

Since in 2012 LinkedIn acquired the professional content sharing platform SlideShare, you have the opportunity to link your Slideshare account and presentation, which is a great way to attract guests.

TWITTER

Twitter is the most popular microblogging platform. "Twitter is based around a very simple, seemingly trivial concept. You say what you're doing in 140 characters or less, and people who are interested in you get those updates. If they're really interested, they get the update as a text message on their cell phone," explains Evan Williams (2013), co-founder of Twitter. In short, Twitter is an Internet-based service that allows people to share short messages (140 characters, called "tweets") to everybody willing to listen. Those who want to listen are people who follow your account on Twitter. So, as in any social network, if you want to use it effectively, you have to start early and set up your networks of followers in Twitter here. The main purpose is to create a solid network of people who follow you. To do this, you can start following other people who you think are interested in your topic. One important term on Twitter is "Twitterfeed", meaning someone's tweets. One of the viral effects of Twitter is to inspire your network to retweet (share) your tweet, thus spreading your message to their own network. In that way your message gets to far more recipients in the form of a recommendation by people who have a good reputation in their own network. In the planning process of your event, you can use Twitter in each step. In fact, we can automatically publish on Twitter from Facebook and LinkedIn. Yet that does not make much sense, as you miss the specific way of communication each network uses. You can use Twitter in the beginning to collect preliminary information. After that, check ideas and topics as part of the planning step.

One of the most popular ways of using Twitter for planned events is during the event itself. People share thoughts, quotes, citations, and opinions during the event. Those tweets reach guests as well as people who are not at your event but are curious to find out what is happening there. Many events use Twitter to collect questions to be answered during a Q&A session. Actually, this is an opportunity that can open up your event to an audience much larger than the guests actually present. Nowadays, it is a must to have a designated person who will send tweets while the event is taking place.

There are two main reasons why event planners use Twitter:

- **Publicity.** Twitter enables the word-of-mouth-effect, thus generating buzz about the upcoming event;

- **Two-way communication with guests.** This is the fastest way to find out about positive and negative experiences, send prompt answers, and maintain a community spirit among your guests and their friends in the networks.

In contrast to Facebook and LinkedIn, on Twitter you have no more than a five-minute tolerance to answer questions, making it very important to plan how you will follow what is happening on Twitter as well as to appoint a person who has the competence and skills to make decisions quickly, who knows the event well, and can manage difficult communication. Never underestimate the power of Twitter to ruin your good reputation. What is more, it is fast and easy to use on mobile devices.

Of course, the principle of starting early is essential too. You have to create an account. It is advisable to turn the "Protect my Twitter updates" off. As in most cases our goal is to gain quality followers, establish relationships, and communicate; however, protecting the updates box is not a good idea. Twitter's appeal lies in its open public nature.

You have to decide how to name your account. It could be named after yourself, the company, or your event. Yet if your event will be happening again in a fixed period of time, it is better to use the event's name. When you do this, you have to start creating your network.

The first step is to listen and choose whom to follow. Also you can search for topics, names, city, etc. Select carefully and think of creating a list of possible ways to divide your network into various topics so as not send all the news to everyone.

Twitter has one simple rule: if you want to answer some tweets, you have to start your message with the name of the person to whom you are responding, for example, @boshnakova. That will show your tweet inmediately on that user's Twitter feed, and he will be able to answer you as quickly as possible.

For an event planner who is using Twitter, it is most important to create a hashtag for the event. It is not always easy to know which keywords to use in order to find the information you need. It is not easy to unite all tweets about your event in one place. At that point hashtags come in handy. "The # symbol, called a hashtag, is used to mark keywords or topics in a Tweet. It was created originally by Twitter users as a way to categorize messages" (Twitter, 2013). You have to use a hashtag to mark all tweets relevant to your event. According to the unwritten laws of Twitter behavior, the hashtag goes at the end of your tweet if it doesn't fit naturally in the text you are writing.

While creating your network, you can search for some hashtags relevant to your business and see who is using them. Before creating the hashtag of your event, make sure it has not been used yet and that it is unique. It is a good idea to use the same hashtag for all social media and networks you use.

When creating your hashtag, you have to spread it among members of your team and your potential guests and encourge them to use it when tweeting about your event. The only thing that you have to do with your hashtag for your event is to let people know about it by using it in all your communication channels: e-mail invitations, Facebook pages, Twitter, and other forms of communication with your attendances. Of all the social networks Twitter is the one that needs the quickest response. So choose carefully when designating who will be responsible for tweeting before, during, and after your account.

Twitter has yet another benefit. Links shared via tweets and retweets will add value to your content and will serve as a search engine optimization (SEO) tool for better search engine results. Think in tweeter style and try to share your most important thoughts within 120 characters. Although Twitter permits 140 characters, it has been proven that 120 charaters give you a better opportunity to be retweeted (RT). And when thinking about your tweet, do not forget about branding, such as by creating a custom background on your Twitter page. This will help remind you that simplicity is always the better choice.

Twitter can be easily integrated into your website, blog, Facebook account, LinkedIn account, and so on. Using, for example, Tweetboard you can bring your event to your site. In that way you, in fact, convert your tweets into an easy-to-follow discussion forum, which is a perfect tool for laying the basis of a real involvement with your followers and website visitors.

Twitter is trying hard to become one of the most utilized platforms in events and has added some new features. Using Twitpic you can share pictures and video on Twitter directly from your phone, which is almost in real time. When creating a new event, you can give it any name or description you like. Once you've added a photo or two to the event, it will have a link you can use to share the entire group of images with the world. Twitpic enables users to organize their images by event, including description, information, and tags for each photo in the group.

Tweet: short message with 140 characters maximum

@person: Public message. If you want to mention somebody, you put @ before her Twitter account name

#name: The hashtag marks a tweet as being relevant to a topic

RT: Retweet is when your tweet is published by another person

FTW: For the win, meaning that's great

NSFW: Not safe for work, which means that the content is not in accordance with the norms of good behavior

Tweetup: A face-to-face meeting organized via Twitter

Figure 10.3. Twitter Terminology

www.socialtoo.com: Tool to auto-follow those who are following you

www.friendorfollow.com: Tool to find out who isn't following you back

www.twittercounter.com: Tool to watch your twitter activity and statistics

www.tweetstats.com: Tool for finding out the patterns of tweets

Project on your wall during your event:

www.twittervision.com: Map of tweets worldwide

Figure 10.4. Twitter Tools

Different options can help you in using Twitter for your events. One of them is TweetMyEvents.com. It is a real-time social media event marketing platform for the social web. With this service you can promote your events to the world through Twitter, Facebook and LinkedIn. Your target market can find about your event and register straightaway. Getting started is easy. Simply sign in with your Twitter account, register your event and tweet it out. There are no limits to the number of events you can display. TweetMyEvents also has a tracking counter, so you can constantly monitor your page views in real time.

In addition, there are other social tools that you should use for your event. Let's look at the latest trends in social media.

FOURSQUARE

Did you check in at Foursquare? If you didn't, please do it now, because you need to share with your friends where you are and what advice you can give them on their arrival there.

Foursquare is a location-based social networking website based on software for mobile devices. Users check in at venues such as restaurants, convention centers, etc. Each check-in awards the user points and sometimes badges. As of September 2013, the company reported it had 40 million registered users and over 4.5 billion check-ins, with millions more every day (Foursquare, 2013). According to one of the co-founders, Dennis Crowley, the company doesn't "want people to think of Foursquare as a game. We want people to think of it as a fun utility. Game dynamics are the bridge to get the people to use it from once a day to three times a day and then two weeks later" (Guardian, 2011).

Although in comparison with other social networks, Foursquare has relatively small user base, the ROI (return on investment) is worth it due to the short-term commitment. Using Foursquare you will have one more tool to enhance audience participation in your event.

To use Foursquare, you first have to set up a new profile that reflects the name and details of your event. Your next step is to set up a venue. That venue will contain again the details of your event. Foursquare allows you to connect your profile with your profiles on Facebook and Twitter. What is very important here is to fill in the proper address of the venue and your contact information. You have various options to choose as category and subcategory, which you should do carefully because potential guests will be looking at Foursquare using keywords, so you have to be sure that everyone interested in your topic will drop by your event. At that point, you can adjust the map to reflect your event location accurately and add tags which will help people be more informed and get the right information at the right time.

On Foursquare people gain badges and became mayors of places they visit frequently. When you get to know the Foursquare network in depth, you will find ways to use badges and mayorships and other instruments to enhance involvement and participation in your event. Of course, if you have a firm budget, you can negotiate with Foursquare and create a custom badge. However, even without a custom badge you can give incentives to your mayor, the people with a high number of check-ins, or those who check in, for example, at most of your places. In addition to tags, you can write tips. Those tips can be used as short messages to your guests and as an instruction on how to get prizes. Why not? It is a good idea to add tips periodically throughout your event and use them to give your guests a basis for frequent check-ins and updates. Think of Foursquare as an instrument to create fun and amusement among your guests. This is a perfect tool for encouraging networking and establishing working relationships, which in most cases is one of the goals of those attending an event. When you start working on your profile here, you will find ways to stimulate participants to add tips too and to share their excitement with friends.

It wouldn't be any news to you if we remind you that it is a must to start early and offer incentives for those who have become your friends early on and can help you spread the news about your event.

As in most cases, your event will be held in a public place where many other events have either been carried out or will take place, and when your event is over, don't forget to delete your tips and "flag your venue as closed." That will open the place for the next event that will be held at the same venue.

Checking areas of your event. One of the reasons event planners use Foursquare is to register the different location of the event as another place and stimulate attendance to register frequently. Planners start by adding tips about the location on the way to the venue where the event will be held, such as airports, railway stations, bus or subway stops, parking areas, main entrances, registration desks, halls, buffets, different exhibition halls, etc. You can take the opportunity to give incentives to guests and in that way encourage them to check in again.

Early on. In most cases you and your main participants will be visiting the event venue many times before the actual event. It is a good idea to ask your collegues and keynote speakers or artists to check in during that period with the intention of reminding their friends that the event is forthcoming. In that way, you will involve participants early on and will use their networks to generate interest and attract attendants. Also during the rehearsal you can give tips to your collegues and encourage them to share their tips with others.

Coordination. The Foursquare platform can be an ideal tool to coordinate your team: if people are asked to check in at any different locations of your event, everyone will know where they are at any moment and can communicate simultaneously with others. The only thing that you have to do in advance is ask all your team members to register and become friends. The rest is easy and fun.

Involvement. At the end of your event you can easily track how many of your guests have checked in, how many of them have shared tips and comments, and which venue or venues were most popular, and collect other important information for the evaluation of the event and for the next event that you will be planning. Don't forget that all these tips and comments get not only to your guests, but to all friends in the network, making it still another means to spread the news about your event.

Sponsorships. This platform opens up new opportunities for engaging guests with your event and sponsors. You can organize games with check-ins at the sponsor's venue, have guests collect badges and post tips or whatever other ideas you can come up with. People like playing games and finding out how to use new technologies. Make the best of people's curiosity.

FLICKR

Flickr is a website used for storing, organizing, and sharing photos. It is also a fun and easy social networking website, popular with photographers and photo lovers. In addition to being a popular site for users to share photographs, Flickr is also a place to manage and organize photos into albums and categories. The world has come a long way in its development, yet it still goes without saying that "a picture is worth a thousand words." As with all social networks, it is important not only to create an account but also to establish your networks.

In your work as an event planner, setting up and maintaining Flickr's account will help you to:

■ tell your event story with the help of pictures

■ reach new potential guests and set up an online community

■ promote your topic and organization with the power of visual communication

■ involve and engage your guests through sharing photos and creating groups

■ acknowledge organizers, sponsors, guests, and partners

As in all other networks, you have a choice between a free and paid account. The free version includes the following: 1 terabyte of photo and video storage; upload of photos of up to 200 MB per photo; upload of 1080p HD videos of up to 1 GB each; video playback of up to 3 minutes each; and upload and download in full original quality and unlimited monthly bandwidth. You can also create groups, which are places to share photos related to a particular theme. Any Flickr member can create or join one, and, generally, they are a great way to share content and start conversations. Groups can either be public, public (invited only), or completely private. Every group has a pool for photos and/or video and a discussion board for talking. There are administrators and members. The other great opportunity offered by Flickr is the Photostream, which allows you to share photos on your web site, blog etc. You are not obliged to upload pictures to all your online places, but just to create links and use all your pictures from your albums in Flickr. Some important things to remember when using Flickr:

■ You can upload six individual photos at a time.

■ Allow time to write titles, tags, and descriptions for your photos. Tags will make it easier for more people to find you. People often tag pictures with names, locations, event descriptions, and themes: "Sofia," "Master class," "Dr. Joe Goldblatt," and "Conference." It will be better if you add some more information to your picture, such as "Conference on 'Social media and networks in event management,' Sofia," for example.

- Integrate your Flickr pictures into your website, blog, or profiles in other social networks.

- Encourage your attendees to do the same. Create a set within your Flickr account to showcase your event photos and share them with everyone. This will help you reach a much wider audience than if you post them only on your website.

- Make sure you have permission to use photos before uploading them. This is particularly important when using photos of children, although that goes for any photos with people in them.

- Consider publishing your photos under the Creative Commons License, which is an open access license with few reuse restrictions. If you do not want your photos to be reused by others, you have to mark the photos when uploading them; otherwise all photos will be licensed as All Rights Reserved. This means that no one else but you can use your photos. The permissions you set will be displayed under each photo. If you want people to share your photos on their blogs or social networking sites, you should choose copyright permissions that will allow them to do this.

- One of the most significant benefits of Flickr is that when you are at an event, the speed of getting up photos online matters. Using the Flickr Uploader tool, you can get your images online very quickly. The closer to your event you can get your photos up, the more likely it is that people will use them to refer to and to share with others, helping to drive traffic.

- Once your imagery starts being noticed, you will most likely receive invitations for permission for your photos to be reused. This means your photos are gaining traction. Try to approve the requests quickly and encourage more people to use your images and credit you accordingly, of course.

- Flickr has an excellent tool that allows you to get deeper metrics on your photos. With these statistics, you can see which of your photos have proved to be the most popular or have been shared with other people, as well as what sites attract people to your photo collections. Don't forget that pictures from the event are essential both for the memories and the engagement of your guests. Upload quality pictures during and after the event; otherwise guests will upload theirs, which are not always of high quality and may not show your event at its best. It is not necessary to upload many photos quickly, although prompt uploading of some of them is obligatory.

YouTube

As far as events are concerned, video content is an indispensable part in our contemporary world. People have gotten used to video as a form of communication; they insist on receiving video content, and they make their own video content and like to share it. Even if you decide not to make your own video about the event, it is possible that your guests will do it. Thus, you had better make your own video, which means you will be in charge and can decide what to show, how to show it, and when to broadcast it. Some numbers:

- More than one billion unique users visit YouTube each month.

- Over six billion hours of video are watched each month on YouTube.

- One hundred hours of video are uploaded to YouTube every minute. (YouTube Insights, 2013)

Just try to imagine what those numbers mean about the way we communicate today. You should consider how to use the video to show your event to its best advantage. It is important to know that 25% of YouTube's total watch time is on mobile devices, so make your video suitable for mobile devices. It is at this point that the effect of the network can be used for your event. People like to share exciting videos; that's why you have to offer them good videos. You can make a video announcing your event, share videos from your event, and use videos from past events to attract visitors to your website or blog, and you may be able to use it for your next event also. Think of the success of TEDTalks, which are videos recorded during TED conferences and TEDx conferences and which attract millions of viewers and generate interest in the conferences. The numbers prove how important it is to communicate via video. According to the site statistics: 50% of the users talk to friends after watching a video and 38% share videos in an additional social network after watching them on YouTube. Video is a powerful tool that event planners should use to reach their goals (YouTube Insights, 2013).

Benefits of using YouTube:

- You can count your audience on the website.

- You can use it to store videos.

- You can make your own channel.

- You can easily share your video with other websites, blogs, social media and social networks.

- You can generate traffic to your event online places.

Some people think that all videos for their event should be made by professionals on YouTube, which is far from the truth. A successful video contains useful information combined with high quality visuals. So think visually and try to broadcast your event's purpose, speakers, atmosphere, and all that you consider helpful in reaching your goals. Videos should be emotionally engaging and need not be necessarily perfect from a productional point of view. All in all, to make successful video content you have to think first about making an emotionally engaging video from the point of view of guests. An involving video is the one that will provide an answer to the most important question: Why should I come to this event? There are plenty of events and only 24 hours a day. So any potential guest will have to choose between your event and a wide range of others. Give the right answer, and they will come to your event. If they miss your event, show them what they have missed. After the event, make them regret not coming on that occasion by making a video from the event, one that will help you maintain your community and will be useful for your next event. However you can, use other opportunities on YouTube.

In your introduction video you can call for action through the pop-up footer banner that appears for 10 seconds into the video and 10 seconds before the end. This banner is clickable and can direct viewers to your website. A video can be a perfect promotional tool, and you should use it to encourage your followers to share it on the web.

You can start your videos before the event—with interviews of speakers, artists, and partners. You can show people what happens before the event or behind the scenes, as well as many other interesting facts. Think as TV producers do; there is always something interesting that's going on. Catch the right moments and share them with the rest of the world.

Your video content is an efficient tool for your PR activities too, thus many bloggers and online media can use it when writing about your event.

A video from your event can help you in two other ways: first, it could be helpful in attracting new sponsors to your next event. You could include a DVD with videos in your sponsorship offers, which in most cases has a much greater impact than written words. That is, in fact, a perfect way to give people an idea of an event. Secondly, your partners and sponsors can use it in their online communication to spread the information to a broader audience.

"Rather than using the standard (and fairly restrictive) YouTube license, the Creative Commons license allows users to offer other people permission to use their videos, as long as they link back to the source. With YouTube's video editor, users can now splice their own videos with others' content, provided that it's available within Creative Commons. These remixes will display an attribution link within the video's 'more' description, but YouTube says that they will make the links more prominent if people aren't clicking on them" (Bazilian, 2011).

Using the "call-to-overlay" function you can combine your video with a link, for example, a link to your external website or to an online event registration website, where participants can buy tickets for your events. The short notice pops up when your video starts to play.

It is great idea to use the "Event Dates" module through which you can list all your event dates and share the complete information of your event on your YouTube channel and give links for tickets online, for example.

To sum up, creating and uploading a video is the first step. As with all networks, you should come up with a relevant and catchy title, add a URL and a clear, engaging Call-to-Action (CTA) to the description box. Consider adding a video overlay that links to your landing page for greater ROI and share it with all your networks, sites, blogs, etc.

SlideShare

In 2012 LinkedIn acquired the professional content-sharing platform SlideShare. That fact created more opportunities for your shared presentations. Slideshare is still the place where people expect to see presentations from your event. However, you can use Slideshare not only after your event: it is a great place to upload interesting information with the goal of attracting people to your event. You can make an exciting presentation of your event, or share past presentations with speakers, making them more popular before the event. There are speakers who upload their presentations before the event chiefly to get feedback from people on what they expect to hear during the actual presentation during the event. Of course, you may think that when you have photos, videos, and text, why do anything else? People are used to presentations. They want to have them or at least access them online. So share your content visually by uploading presentations before and after the event.

| 60 million monthly visitors |
| 130 million pageviews |
| Amongst the most visited 200 websites in the world |

Figure 10.5. SlideShare Quick Facts (Source: http://www.slideshare.net/about 27.12.2013).

Another reason to do this is that you can reach people who, otherwise, will not visit your website, blog, or profile in social networks. You can use LeadShare, which allows you to leverage your content by capturing the viewer's contact information (http://www.slideshare.net/SlideShareHelp/leadshare) or Send Tracker, which allows you to send a link to a contact and get notified when that person opens the e-mail and clicks on the presentation. It highlights a summary of all activities and provides links to detailed reports on each generated tracker. However, those services are paid services.

Now it is possibe to syncronize MP3 audio with slides to create a webinar. This is a great opportunity to present your events, the speakers, the topic, etc. It is free and you need little time to get used to do it. The next opportunity is to embed YouTube videos inside presentations uploaded on Slideshare. Of course, there are other paid options, such as branded channels, analytics, ad free pages, and many more.

Google+

At the end of 2013 G+ was not far behind Facebook at just over 50% users globally. We have to keep in mind that a Google+ account is mandatory whenever you create a new Gmail account. This has been pushing up the account ownership stats. No other social network has Google's web assets leverage. Besides, there were many new features keeping the social networks alive and competitive with the marketplace. Google has been integrated into almost all of the other Google web services. In fact, that is the situation with events. The idea is to make it easy to schedule events and keep track of details (attendees, photos, etc.). Among the benefits worth mentioning are the following: (http://www.google.com/+/learnmore/events/#):

- All your Google+ Events appear in your Google Calendar. When replying or creating an event in Google+, you can check your Google Calendar schedule right from the event to see if you have a conflict.

- Easy e-mail reminders are automated reminders that are sent out to your confirmed guest list, along with an update when the event is starting.

- With Google+ Events, everybody attending the event can share their photos in a single photo collection; as a result, photos aren't scattered across different websites, albums, and e-mail attachments as they used to be in the past.

- Party Mode allows you to share pictures and videos you take during an event instantly with everyone at the event, and with the Google+ Android app installed, you'll have the option to turn on Party Mode when an event you're attending begins.

- Get the right look of the occasion with animated event themes, or give your invite a personal touch by adding a picture from your phone, computer, or one of your Google+ albums.

With "Google+ Hangout" you can plan a group video call with your partners by selecting the option when you create an event. Everyone will be reminded to join the Hangout when the scheduled time comes around and you can talk face to face. However, it is possible to use it for marketing or take advantage of the "On Air" option to broadcast your Hangout publicly and record it for sharing later.

Actually, the influence of Google+ on search results is serious enough to prove that it is a platform one cannot afford to ignore. With just a few clicks you can fill in the information about your event and link to the event registration page where attendees can purchase tickets. And last but not least, you can use Social Analytics to gain insight into offsite activities involved in your Google+ Event. That information is very helpful for your next event, as you can see which activity and content had the greatest impact.

Instagram

The sale of Instagram to Facebook in 2012 was the ultimate Silicon Valley fairy tale. By 2013 the Instagram community had grown to over 150 million people. That same year the network introduced the opportunity to share video. Using the mobile application you can take up to fifteen seconds of video and share it with your friends. When you post a video, you can select your favorite scene as your cover image, so your videos can be appealing even when they're not being played. From Instagram you can share your photos and videos easily with Facebook, Twitter, Tumblr, etc. Moreover, celebrities and brands quickly latched onto the new feature, such as pop star Justin Bieber who became the first Instagram user to earn one million "likes" on a video post (Hernandez, 2013). We are well aware of the fact that people like images, and the ways we can use Instagram for our events are boundless, limited only by our imagination. However, one must never ignore the importance of the hashtag, which makes all our photos and videos accessible to other people, just as our guests can share their own pictures with us and the whole community, thanks to the hashtag. Many people prefer to share their emotions through Instagram because it is fast and user-friendly.

In practice event organizers use Instagram for:

- creating anticipation: by photos, quotes, short video announcements

- sharing about the event: real-time information from the room

- sharing behind the scenes: everyone is interested in what is happening out of sight

- organizing competitions: guests like to participate in fun activities, especially if they are easy and quick

- follow-up comments: short videos from participants.

As with all social media and networks, you should communicate effectively and promptly on the hashtag of your event.

If you want to go deeper into Instagram, you can have a look at printstagr.am, a website where you can print selected pictures from Instagram. Also, Eventstagr.am is a photo curation platform with which you can curate event photos shared on the Instagram social network in real time quickly and easily. Fully customizable, Eventstagr.am collects and displays event photos based on the use of an event-wide hashtag. Participants simply tag their Instagram photos on this hashtag, and Eventstagr.am is activated, collecting and curating user-submitted photos and videos into a clean multimedia display. In fact, you can display the photo stream using a projector or TV screen and watch.

Pinterest

When you open the website, you will read "Pinterest is a tool for collecting and organizing things you love." That sounds great for any event. Pinterest has over 70 million users and is steadily gaining momentum outside the U.S. Another interesting fact worth mentioning is that nearly 80% of users are women, and 35% access it only from mobile devices (Smith, 2013). Being one of the latest platforms, the ways we can use it for events are still to be discovered. Meanwhile, we can share some of those which work well in practice. Pinterest is a convenient tool for

- collecting ideas for your next event: There are a lot of pinboards with pictures from events, publications, and other inspirational information.

- crowsourcing ideas for events: You just need to create a pinboard and invite people to share their insights with you.

- selling your event to guests, sponsors, and stakeholders: Visualization is always a great way to present your ideas and prospects.

- collecting and sharing the content from your events by means of an event organizer or by guests invited to pin photos with highlights of the event.

- creating a place for follow-up: collecting all links with feedback from the event in one place.

- generating traffic to your website, blog or other content online.

Of course, we can use it for self-promotion, as it offers great visual opportunities as well as sharing links. It is a good idea to pin all those opportunities and look for more.

MySpace

If you plan a social or pop culture event, it will not be reasonable to skip MySpace as a platform. Just as with all other social media and social networks, you have to start early on and create your network. In the middle of 2013 Justin Timberlake, who acquired, together with his partners, Chris and Tim Vanderhook, the site for $35 million,

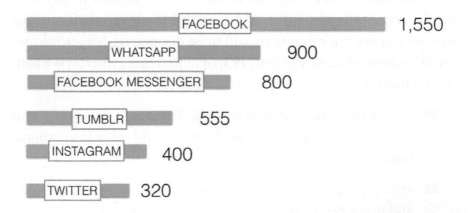

FACEBOOK 1,550
WHATSAPP 900
FACEBOOK MESSENGER 800
TUMBLR 555
INSTAGRAM 400
TWITTER 320

*Number of active users in million as of January 2016

Figure 10.6. Number of Active Users of Socia Networks Worldwide (data from http://www.statista.com)

introduced a revamped site with a musical focus. Although the new design does not have an application for events yet, it is forthcoming.

The site has been redesigned to focus exclusively on music, so if you are into music events, you have to watch out for new developments on MySpace to grab the new features about events as soon as they are introduced.

As we have already mentioned, it is not possible to make a comprehensive analysis of all social media and networks. Not being able to go deep into details, we are aware that changes are underway every single day and that technology is evolving so quickly that we can hardly keep pace and manage to study it. We cannot but agree with Sir Ken Robinson who said, "It is an interesting feature of cultural change that for a period of time, new technologies tend to be used to do the same old thing" (Robinson, 2001). What we can do is to watch carefully for all changes and try to use them creatively to make our guests' experience better.

How to measure your success in social media and social networks

It goes without saying that measuring your sucess in social media is top priority. Although social media and social networks may look like free communication channels, in fact they require a lot of time, creative ideas, and dedication. So anyone who uses them for business purposes is interested to see the results. Those channels are still very young, and it is difficult to talk about standards in social media and social network measurement and evaluation. However, their digital nature facilitates checking progress in achieving one's goals. Figure 10.7 shows the top 20 free tools by which you can measure and evaluate your work. Be prepared to spend enough time on evaluation because that is the only way to be sure what to do in social media and networks for your next event. It has already been mentioned that those tools are not fixed; new tools will emerge as specialists are working out the standards in the measurement of those activities.

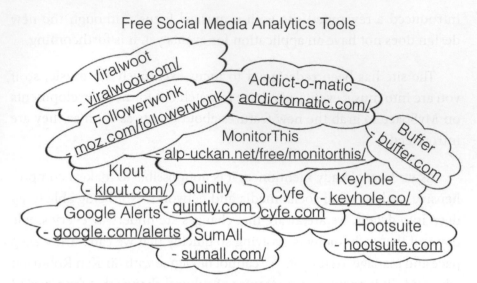

Figure 10.7. Free Social Media Analytics Tools

One important aspect of social media and social networks that must be kept in mind is that people visit those channels because they need to communicate; they want a conversation with you as presentor of your event. Real success means not just starting a conversation but being able to maintain it as well. It is only then that you have achieved a real long-lasting success, which, however, must be measured and defined by your objectives. So from the very beginning you have to be convinced in what exactly you want to achieve for your event by using social media and social networks. In the end we would like to finish with a quotation by Albert Einstein, who said, "Not everything that counts can be counted, and not everything that can be counted counts." It is necessary to measure and evaluate your success in social media and social networks, but you also have to remember that the most valuable asset is the community you create, and your success will depend on your ability to keep that community alive until your next event, repeating it again and again after each planned event you have done.

SUMMARY AND CONCLUSION

Since 2004, when Facebook was initially launched, social media and networks have not only transformed the way we communicate, but also the way we make decisions and spend our time with friends, the way we do business, and last but not least, the way we organize special events. We know for certain that those changes are here to stay and more will follow. We cannot possibly foresee what will come next, but we can accept the changing paradigm and the fact that if we want to be good in what we are doing, we have to change every single day. It is not possible to be in all social sites, and it is not necessary. What we can be certain of is that people are using mobile devices more and more and that they prefer video content and want everything to happen fast and to be online. We can adapt to that situation and anticipate the next challenge.

DISCUSSION QUESTIONS

1. How do social media and social networks change the event experience?

2. Are social media and social networks simplifying event planning or is it just the opposite?

3. How does the new media help local events to come to the global scene?

TASKS

1. Read the article at http://www.theguardian.com/uk/2011/mar/02/royal-wedding-website-launched to see how event planners have used new technologies for making the wedding a global event. Analyze how they have used social media and social networks and how those new ways of communication have changed the event industry. Try to make a forecast with the top five changes that will take place in the field of planned events. Compare your forecast with those of your colleagues.

2. You are organizing a major art festival that will be featuring artists from around the world. It is a one-week festival that will include more than 300 artists from 40 different countries. You expect to have more than 60,000 guests. Some of the shows will be conducted in the open, on the streets, or other public places.

Describe which social media and networks you will use and why. Be specific about your objectives with each one.

REFERENCES

Adler, Emily. 2014. Social Media Engagement: The Surprising Facts About How Much Time People Spend On The Major Social Networks, accessed 9/10/2015

Amiando Event Registration and Ticketing. 2011. Social Media & Events Report 2011: How Is The Event Industry Using Social Networks? accessed 9/10/2014. https://www.amiando.com/fileadmin/Data/Info-Center/Reports/Social_Media_Report/Social_Media_and_Events_Report_2011.pdf

Backstrom, Lars, 2011. Anatomy of Facebook, accessed 26.12.2013

Bazilian, Emma. 2011. YouTube's Creative Commons License Makes Mash-Ups Easy Remixed videos will link back to the source, accessed 27/12/2013

Bullas, Jeff. 2013. "12 Awesome Social Media Facts and Statistics for 2013." http://www.jeffbullas.com/2013/09/20/12-awesome-social-media-facts-and-statistics-for-2013/Halliday, Josh, Dennis Crowley: Don't think of Foursquare as a game

Cohen, Lon S. 2009 . "Is There A Difference Between Social Media And Social Networking?" http://lonscohen.com/blog/2009/04/ accessed 9/10/2015.

eBizMBA Rank. 2015. "Top 15 Most Popular Social Networking Sites—September 2015." Accessed 9/10/2015. http://www.ebizmba.com/articles/social-networking-websites

Facebook Newsroom, Events, accessed 9/9/2015. https://newsroom.fb.com/Products/

Foursquare. 2013. About Foursquare, accessed 26/12/2013.

Gossieaux, Francois and Ed Moran, The Hyper-Social Organization: Eclipse Your Competition by Leveraging Social Media, New York, NY: McGraw-Hill, 2010

The Guardian, Monday 4 April 2011, accessed 26/12/2013

Hartshorn, Sarah. 2010. "5 Differences between Social Media and Social Networking," Social Media Today. www.socialmediatoday.com/content/5-differences-between-social-media-and-social-networking. accessed 26.12.2013

Hernandez, Brian Anthony. 2013. "First Instagram Video To Reach 1 Million Likes Stars Shirtless Bieber." Accessed 9/10/2015. http://mashable.com/2013/06/22/bieber-instagram-video/#4i7V0J8OIOkN

Kawasaki, Guy. 2011. "Ten Ways to Use LinkedIn." accessed 09/06/2011

Robinson, Ken. 2001. Out of Our Minds: Learning to be Creative, 2001. Capstone; March 15, 2001)

Smith, Craig. 2013. "36 Amazing Pinterest Stats." accessed 27/12/2013

Social Media & Events Report 2011", accessed 26/12/2013

The Steveology Blog. 2012. "Top 20 Free Social Media Monitoring Tools." http://www.slideshare.net/stevefarnsworth/top-20-free-social-media-monitoring-tools-from-the-steveology-blog-2013?qid=cfbfc2b7-2d62-47fa-b4f7-758fd94cf2d5&v=default&b=&from_search=1 Twitter. 2013. Using hashtags on Twitter, accessed 26/12/2013

Visually. 2013. Accessed 9/10/2015. shared by forwardsolutionsph on Sep 05, 2013 in Social Media. http://visual.ly/top-10-most-popular-social-networks-2013

Williams, Evan, The voices of Twitter users, accessed 26/12/2013

YouTube Insights. July 2013, http://www.google.com/think/research-studies/youtube-video-insights-stats-data-trends.html. Accessed 27/12/2013

CHAPTER 11

Mobile Applications for Meeting and Event Marketing

> *"If your plans don't include mobile, your plans are not finished."*
>
> —*Wendy Clark, Coca Cola (McLandress, nd)*

LEARNING OUTCOMES

As a result of reading this chapter, you will learn how to:

- Apply fundamentals of marketing to mobile meetings and event marketing and identify difference in mobile application-based marketing from traditional marketing

- Identify the key players in the mobile event marketing application

- Develop a basic mobile marketing plan by applying mobile applications

- Recognize the business opportunities in the mobile marketing application market

- Evaluate the existing mobile event marketing apps

INTRODUCTION

It is never an overstatement to say that "mobile applications" are the present of technology in many businesses and many countries. The photo below is representative of the adoption of mobile devices by people with all levels of technological background. The meeting and event industry is not an exception to this statement. The explosion of affordable and powerful smartphones with faster wireless broadband is accelerating the mobile applications market. Consumers are getting used to improved capability of their devices through many useful apps. Current mobile apps provide enhanced features plus mobility, which once only desktop-based applications could have provided.

It is important to note that, actually, software developed for mobile phones had been around for over a decade before the term "app" was popularized by general consumers. The Apple App Store was considered the first app store; it allowed people to search and purchase applications easily, and it was the first to provide a platform for app developers to sell their applications with a simple process.

Photo 11.1. A Kenyan Farmer Uses a Mobile Phone in the Field. (Photo by Neil Palmer; used with permission via the Creative Commons Attribution-Share Alike 2.0 Generic license.)

It was recorded that 10 million apps were downloaded within its first week of operation. Android and iOS (Apple) have their own app stores, and other smartphone manufacturers are also operating their own application stores now. In 2015, it is projected that smartphones will constitute over 70 percent of the overall sales in mobile handset. According to a research report from the analyst firm Berg Insight, the number of mobile application downloads wordwide reached 60.1 billion and annual downloads will reach 108 billion by 2017. It is expected that iOS and Android combined will serve over 62 percent of total app downloads. Windows phone operating system is projected to be the third most popular application platform in 2015 (Berg Insight, 2013).

Mobile application is a global phenomenon. The report estimates that revenues from paid applications and in-app ad revenues reached € 6.4 billion in Europe in 2012 and expects it to grow €14.1 billion at an annual growth rate of 17 percent. Third-party app stores are especially popular in China and other markets where Google Play hasn't become the default on-device app store The Android platform, the second largest platform, generated € 80 million in 2010. It is forecasted to grow to € 1.5 billion in 2015. The following Figure 11.1 shows the changes in mobile apps download over years.

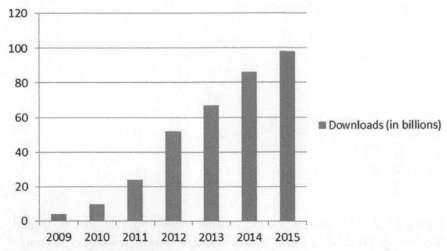

Figure 11.1. Mobile Application Downloads (Source: Berg Insight 2011).

UNDERSTANDING NEW PARADIGMS IN TECHNOLOGY-BASED MARKETING

Internet Marketing, E-Marketing, and Digital Marketing: What Is the difference?

Many business professionals especially, small business owners such as event management organizations, hear about these new technology-based marketing tools. However, many of them do not have formal education in marketing or technology-enhanced marketing. Answering this question and helping meeting and event planners to gain knowledge should be approached to explain the scope of each marketing tool and how to make the most of each tool to maximize marketing outcomes.

Various terms related to technology-based marketing have been used. They include Internet marketing, digital marketing, and mobile marketing. According to the Google Trends report (2015) by search terms (Figure 11.2) , there is a clear change over the last decade. Internet marketing was the most searched term till 2013, but digital marketing took over the spot, and the number of times it has been searched has drastically increased since then. Therefore it is critical to understand each concept and scope.

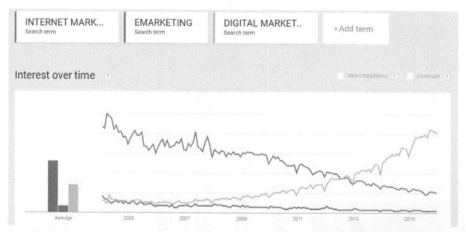

Figure 11.2. Web Search Interest: Internet Marketing, Emarketing, Digital Marketing: Worldwide (Data Source: Google Trends [www.google.com/trends]).

Internet marketing is defined as advertising and marketing efforts that use the Web and e-mail to drive direct sales via electronic commerce by Beal (n.d.). Internet marketing includes specialized areas such as Web marketing, e-mail marketing and social media marketing (Beal, n.d.). Internet marketing also used to be listed in top in Google search, known as search engine marketing. However, the scope of Internet marketing does not take full advantage of broader and more powerful digital media. Therefore, for Internet marketing to be successful, it is recommended that it be integrated with digital media as well as with traditional media such as print, TV, and direct mail.

E-marketing may be often interchangeably used with Internet marketing and digital marketing in many industries. However, e-marketing is considered to have a broader scope than Internet marketing since it utilizes various digital media, including website, e-mail, and wireless media. It also provides digital customer data and electronic customer relationship management systems (Chaffey, 2013). These techniques are used to acquire new customers and provide requested services to existing customers to achieve customer relationships. Today's customers are multigenerational and have distinctive differences in the ways they obtain product information. Therefore, it is very critical to combine market channels that result in acquiring new customers and retaining existing customers. This is the most effective marketing system: to use a combination of digital and traditional channels in customer communication and product distributions. This is called multi-channel marketing, and its concept is well summarized in the following Figure 11.3, developed by Chaffey (2013).

Digital marketing is a term increasingly used by specialist digital marketing agencies and the new media trade publications. To help explain the scope and approaches used for digital marketing and how digital marketing needs to be closely aligned to broader marketing objectives and activities, details of digital marketing are shared here. Digital marketing is a term that has gained popularity since 2013, according to the Google trend search (as seen in Figure

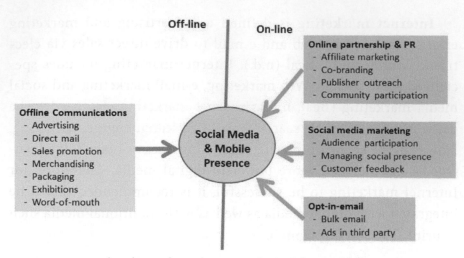

Figure 11.3. Multi-Channel Marketing (Adapted from Chaffey, 2013)

11.2). While there are multiple definitions from different sources, a simple version of its definition is probably this one, from the Digital Marketing Institute (http://digitalmarketinginstitute.com): "Digital marketing is the use of digital channels to promote or market product and services to consumers and businesses." Due to its use of the general term "digital channels" and scope of its channels, digital marketing is considered an umbrella term for many types of technology-based marketing. Its utilized digital channels include e-mail marketing, search marketing, social media, online PR, video advertising, infographic, affiliate marketing, and display advertising (Chaffey, 2013). No matter which specific channel or a combination of multiples is utilized, it is critical to adopt a targeted, measurable, and interactive approach.

As digital marketing involves many online channels (web, e-mail, database, and mobile/wireless devices), still many consumers or small businesses interchangeably use it with online marketing. Chaffey (2013) argued that digital marketing is a shift from product-focused marketing to customer-centric marketing because it utilizes tailored combination of channels based on customers' profiles created by research. As a result, digital marketing can provide more effective marketing tools to meet customers' needs.

MOBILE MARKETING (M-MARKETING)

Mobile marketing is evolving fast and is rapidly becoming a major communications channel for reaching today's customers. Its growth is fueled by the explosive increase in the number of mobile users and the number has already surpassed the number of global laptop users in 2014 (See Fig. 11.4). Mobile marketing utilizes mobile devices as the channel of marketing. According to Mobile Marketing Association, a broader scope of mobile marketing includes advertising, apps, messaging, m-commerce, and customer relationship management on all mobile devices, including smartphones and tablets (http://www.mmaglobal.com/about). Due to the nature of mobile phones—the most direct and personal channel—mobile marketing is often considered consumer-initiated because it requires the consent of consumers to receive future communications.

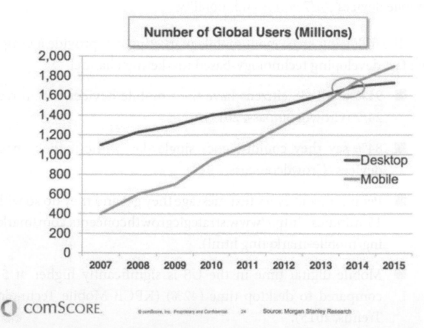

Figure 11.4. Number of Global DeskTop and Mobile Users (Source: SmartInsight, 2015)

Figure 11.5. Mobile Poll (Source: Crowdcompass, 2012)

In addition, m-marketing can provide customers with time-specific and location-based information that can lead to impulsive purchases due to mobile devices' 24/7 access and mobility.

The following facts about mobile marketing can provide a grasp of the fast developing technology-based marketing channel.

- 91% of all US citizens have their mobile devices within reach 24/7 (Crowdcompass, 2012).

- 84% say they couldn't goes single day without their mobile devices (Crowdcompass, 2012)

- People look at every text message they get and they do so within 15 minutes (http://www.strategicgrowthconcepts.com/marketing/mobile-marketing.html).

- Mobile digital time in the US is significantly higher at 51% compared to desktop time (42%) (KPCB Mobile Technology Trends, 2015).

- 80% of Internet users own a smartphone (SmartInsights, 2015).

■ Within five years, half of today's smartphone users will be using mobile wallets as their preferred payments method (Carlisle & Gallagher Consulting Group, 2012)

■ 25% of international media and marketing executives see mobile as the most disruptive force in their industry (AdMedia Partners, 2013)

Significant advantages of using mobile-marketing include its personal, meaningful, and long-lasting communication with customers that leads to being engaged with brands and improving relationship with customers (Powell, 2010).

In summary, the question of why companies have to consider mobile marketing can be answered as follows:

1) Internet is mostly used via mobile: According to Forrester Research World Mobile and Smartphone Adoption Forecast (2014), about a fifth of the world's mobile subscribers are currently using their mobile handsets to go online, and mobile Internet users will exceed PC-based Internet users in 2016. Low-cost data plans will allow many consumers who have not been able to go online to access the Internet.

2) Mobile marketing is effective: While mobile marketing won't replace all other marketing channels, it enhances other channels. Combined with traditional channels (print, TV, radio, direct mail, e-mail) and other channels, mobile marketing will expand the range and track the effectiveness of those channels.

Mobile marketing is most effective when utilized as a complement to the marketing mix in conjunction with your standard marketing media.

CHALLENGES FACING M-MARKETING

Concern about the security of sharing financial information over a mobile device was reported as one of drawbacks, and about 25% of

Figure 11.6. Reasons of Resistance to the M-Commerce Experience (Source: UPS Pulse of Online Shopper, 2014).

respondents said that the check-out process via mobile is too bothersome (UPS Pulse of Online Shopper, 2014). Figure 11.6 shows the reasons that e-commerce shoppers are still resistant to the m-commerce experience. It is reported that more than 50% of consumers prefer online shopping, but they haven't converted to mobile purchasing yet. Almost half of the consumers surveyed said they will forego making a purchase on their mobile device and use a PC instead, because the product image isn't large or clear enough. About a third of the respondents said product information isn't large enough on mobile shopping apps, and 3 in 10 said comparing products was frustrating on a mobile device. It is important to note that m-marketing can be a stand-alone marketing channel; however, a better chance of maximum outcome exists when m-marketing is utilized as one channel of a multi-channel marketing approach as it will be better suited to customers' individual needs.

TRENDS IN MEETING AND EVENT MANAGEMENT MOBILE APPLICATIONS

As discussed earlier in this chapter, marketing has been transformed by the Internet and related technologies. This transformation has allowed marketers to move from product-centric to customer-centric marketing. Therefore, the meeting and event industry is also shifting its marketing focus from offline advertising to personalize digital mobile marketing. Several chapters in this book have touched on the adoption of mobile apps in many different event and meeting management functions, including event floor design, registration, virtual meetings, and more. More and more meeting and event attendees are adopting event mobile applications because of their convenience and mobility. With wireless broadband enabling them, many mobile applications are being developed using mobile phones and other handheld devices, which will have direct meeting applications. For example EventMobi™ has moved all event management functions into mobile apps. It allows both meeting and event planners to access key information in real time with the convenience of mobility during attendance meetings and events. Log-On (www.log-on.nl) uses mobile phones to provide a range of tradeshow and event applications for attendees, including conference agendas, exhibit product directories, event feedback and surveys, SMS audience polling, group announcements, networking capabilities, travel information, and more.

The meetings and events industry was highly social, mobile, and collaborative in its nature even before the mobile and social media technology began to rule the business world. These characteristics are ideally suited for the benefit from m-marketing. The use of wireless media provides an array of technologies to choose from, but the always-on, always-connected personal communication channel mobile devices can be a game changer in the meeting and event industry. To event and meeting professionals, the mobile device can be a new and powerful channel to get in touch with event prospects as it is in their palms most of the time; some people have their smartphones next to them 24/7. However, many meeting and event professionals do not possess advanced knowledge or the skill sets to develop mobile marketing strategy or create mobile marketing apps yet. Creating and managing a mobile-marketing strategy may seem overwhelming for meeting and event planners who are already responsible for many planning responsibilities, especially if a firm is a small business and has no marketing or mobile technology expert available. The good news is that cloud and mobile innovations can help even the most inexperienced meeting and event professionals implement a successful mobile marketing strategy on an affordable budget. Therefore, it is suggested to benchmark other business disciplines and how they adopt and implement m-marketing to increase their sales and awareness of their products and services.

Practical m-marketing strategy for meeting and event organizations

Getting the prospective attendees to visit your event site is a first step. The event page should be engaging enough to make a reservation or at least to revisit easily for more information later. Meeting and event professionals often make the mistake of not considering the uniqueness of mobile phones. For example, information developed for PC viewing may use fonts that are too small or which have too much text for viewing on mobile devices. It is critical to consider the mobile user who has a small screen. Here are some tips for developing appropriate content for mobile devices:

- Start with analyzing your website. Some of the key areas to consider include a) the text: If the text heavy, it may strain mobile users' eyes and patience, b) buttons may be hard to find or too small for mobile users to use, and c) event information should be easy to find so attendees aren't frustrated with the search.

  Upon these evaluations make necessary adjustments to accommodate prospective attendees who will access meeting and event information via their mobile devices and make reservations.

- Think small screen first for e-mail: E-mails are a powerful tool that can increase revenue by 28 percent. They also include powerful analytics, so meeting and event professionals can understand which marketing campaigns are working and why. Much like the website development tips above, consider small screen first, with easy-to-click links.

■ Think of blogs as your mobile-chat platform: Blogs are a great way to connect with attendees before, during, and after an event or meeting, engage in conversations, and offer relevant information. Seventy-one percent of people read blogs on their mobile devices, so make blogs mobile friendly. An additional benefit of readable blogs is that they can even give your search engine optimization (SEO) strategy a boost, as search is prioritized by readability.

■ Adopt a holistic mobile strategy: Only nine percent of the top online retailers are using responsive design, which ensures business messages and accompanying fonts, graphics, buttons, and links are designed for small screens. Successful businesses will adopt strategies that specifically cater to mobile users.

■ Pivot faster with the help of the cloud: Cloud can help meeting and event organizers create marketing campaigns, conduct testing to determine effectiveness, make changes in subject lines, and alter websites almost instantly. M-marketing can be especially benefited by cloud technology: responsive-design templates "made for the mobile screen" are built into products, from e-mail marketing software to web design software. This flexibility and adaptability will help event and meeting professionals be updated with the fast advancing mobile technology and meet planners and attendees' requests.

The mobile market is a terrific opportunity for many event and meeting organizers. The main focus should be on how to plan your event content according to what prospective attendees consider important. The event content should be presented in a variety of forms. With the right tools toward the right target attendees, meeting and event organizers can realize an immense attendee increase and revenue boost plus establish ongoing connections with potential repeating attendees.

What is in m-marketing for the meeting and event industry?

It is a fact that the details of an event and a meeting can often change until the event is over, so finding ways to deliver and attract people's attention over time is critical in event marketing. An event marketer's dream must be to constantly engage with prospective attendees in meaningful two-way conversations. Mobile is close to being the dream marketing channel because it provide a direct line of communication with customers. For attendees mobile provides them access to events at any time, wherever they are. M-marketing can help planners reach prospective attendees via their mobile phones/smartphones—which are now the most used tools to research and search information for products and services, including via text message, Smartcode (2D), proximity Bluetooth advertising, and location-based social media (LBSM).

Corbin Ball, who is a meeting and event technology consultant, has projected "Twelve Technology Trends in the Meetings and Events Market" (Ball, 2014). He pointed out that event marketers will have the ability to market inexpensively to targeted audiences using multi-media based blogs (online journals) and RSS (really simple syndication) newsfeeds.

SUMMARY AND CONCLUSION

Mobile apps have deeply shifted the way customers in the meeting and event industry gather information, learn about events, and participate as attendees. Event marketers and planners need to effectively leverage mobile event apps to elevate the event experience—from the moment event registration opens to the last attendee returning home. Mobile marketing will add innovative and effective marketing for conferences, meetings, conventions, trade shows, and events with appropriate app development, which will result in a better event experience. No longer just a trend, this new standard—mobile marketing—means that planners need to be aware of mobile applications and how to develop mobile event marketing to reach and attract attendees. Event marketing requires a holistic digital marketing approach that includes e-mail, website, and mobile marketing with consideration of design. To help event planners make sure that mobile marketing is aligned with the growing meeting business, it is important that they develop a marketing strategy that includes how to achieve growth through e-marketing.

Employment in advertising, marketing, and public relations is expected to grow 13 percent through 2022, according to the Bureau of Labor Statistics (1984). For students who are studying meeting and event management, marketing skills in mobile marketing can distinguish themselves as this skill is now recognized and will provide competitive advantages for their career and for their employers. They are in high demand in the meeting and event industry. Employers will seek those individuals who have the appropriate skills to conduct new types of marketing and public relations involving new media, particularly mobile apps.

DISCUSSION QUESTIONS

1. What types of mobile marketing tools/contents have you experienced in the meeting and event industry?

2. What are the unique aspects of mobile marketing? List the differences from traditional marketing

3. How do you (or your organization) align mobile marketing strategy with your overall business objective?

4. How can meeting and event planners benefit from mobile applications?

5. Which strategies will be the most important for mobile app-based event and meeting marketing?

6. Why is it important to implement mobile app-based event and meeting marketing?

7. How can mobile app marketing increase revenue and the number of meeting and event attendees?

TASK

Discuss a tailored event marketing plan for various age groups (baby boomers, Generation X, and millennials [Generation Y]). Identify what the most effective mobile marketing contents for each group are. List potential challenges of implementing mobile marketing for various types of events (e.g. conference, special event, tradeshow).

REFERENCES

AdMedia Partners (2013). Mobile Perceived as Most Disruptive Media and Marketing Trend. Retrieved on July 11 2015 from http://www.media-post.com/publications/article/191924/mobile-perceived-as-most-disruptive-media-and-mark.html?edition=

Ball, Corbin (2014). Tech Talk Newsletter. http://www.corbinball.com/techtalk

Beal, Vangie (n.d.). Retrieved on November 11, 2014 from http://www.webopedia.com/TERM/I/internet_marketing.html

Berg Insight (2013) The Mobile Application Market, 2nd edition by Berg Insight (2013). Retrieved on December 21, 2014 from http://www.prnewswire.com/news-releases/the-mobile-application-market---2nd-edition-108-billion-mobile-applications-will-be-downloaded-in-2017-210410311.html

Chaffey, Dave. (2013) Definitions of Emarketing vs Internet vs Digital marketing. Reviewed: 24, May 2014. Accessed: http://www.smartinsights.com/digital-marketing-strategy/online-marketing-mix/definitions-of-emarketing-vs-internet-vs-digital-marketing/

Crowdcompass (2012). "Mobile Apps: The New Events Essential." Retrieved on November 22, 2015 from http://www.crowdcompass.com/pdf/mobile-apps-the-new-event-essential.pdf

Forrester Research World Mobile and Smartphone Adoption Forecast, 2014 Google Trends. (2015). Retrieved on December 10, 2015 from http://www.google.com/trends/explore#q=INTERNET%20MARKETING%2C%20EMARKETING%20%2C%20DIGITAL%20MARKETING .

KPCB Mobile Technology Trends (2015). http://www.kpcb.com/internet-trends.

McLandress, Krista (n.d). Top 10 mobile marketing quotes (from the pros!), http://blog.apps-builder.com/10-mobile-marketing-quotes/

Mobile Marketing Association, Accessed: (http://www.mmaglobal.com/about).

Mobile Marketing Tips. Sharing best practices to help your business grow. Reviewed: 23, May 2014. Accessed: file:///C:/b-research/2013%20publication_4/Book_Meeting%20and%20Event%20Technology%20chapter%20Nov%2020/ch%2011%20mobile%20app/Mobile_Media_12-11aSP%20ONLINE%20VERSION.pdf

Powell, Frank (2010). "Five Advantages of Mobile Marketing over Online." Mobile Marketer. Retrieved on April 11, 2015 from http://www.mobile-marketer.com/cms/opinion/columns/5478.html

SmartSight (2015). Mobile Marketing Statistics 2015. Retrieved on July 29 2015 from http://www.smartinsights.com/blog/

UPS Pulse of Online Shopper (2014). https://www.ups.com/media/en/2014-UPS-Pulse-of-the-Online-Shopper.pdf

CHAPTER 12

Guest-Generated Content

> "Consumer generated content is rapidly gaining traction as part of the purchase decision making process."
>
> —Peter O'Connor, Institute de Management, Hotele International, Essec Business School, France (2008)
>
> "Once upon a time there were two countries, at war with each other. In order to make peace after many years of conflict, they decided to build a bridge across the ocean. But because they never learned each other's language properly, they could never agree on the details, so the two halves of the bridge they started to build never met. To this day the bridge extends far into the ocean from both sides, and simply ends half way, miles in the wrong direction from the meeting point. And the two countries are still at war."
>
> —Vera Nazarian, The Perpetual Calendar of Inspiration (2010)

LEARNING OUTCOMES

As a result of reading this chapter, you will learn how to:

- Encourage and drive guest-generated content to your meeting and event website and through social media

- Moderate guest-generated content for your meeting and event to ensure appropriate comments are included

- Promote democracy and transparency in terms of guest-related content

- Avoid bullying and harmful comments through guest-related content

- Measure and evaluate guest-related content for evaluative purposes

INTRODUCTION

Each day tens of millions of guest-generated comments are added to the World Wide Web. These comments may be seen through consumer media such as online newspapers, blogs, social media, and other electronic channels. When Sir Tim Berners-Lee helped weave the worldwide web, he submitted guest-generated comment when he wrote "This is for everyone" (Berners-Lee, 1999).

Through the development of the World Wide Web, both opportunities and challenges have risen regarding developing and managing guest-generated content. For the purposes of this chapter, the terms guest-generated and user-generated content will be used interchangeably.

Guest-generated content is a critical tool for meeting and event professionals in that it promotes engagement in the content of the meeting or event as well as widens and deepens the relationship between the guest and the host organization.

Historically, once the flood gates of the World Wide Web were first opened in the early 1990s, the concept of guest-generated content merely naturally and exponentially evolved without a specific strategy for cultivation, conversion, and evaluation of the engagement of the individuals submitting their comments and opinions.

According to the Book of Genesis in the Old Testament, the Tower of Babel represented a single language that united humanity following the great flood. In the twenty-first century we now have a similar flood of information and must develop a language to enable our meeting and event guests to effectively engage with one another as well as the content of our programs.

ORIGINS, MOTIVATIONS AND INCENTIVES OF USER-GENERATED CONTENT

According to Levine and Prietula (2013), user-generated content is often produced through open collaboration. The term first emerged in 2005 through web publishing and new media content. Websites such as Amazon.com and TripAdvisor.com rely on user-generated content to provide reviews and promote sales. The concept of user-generated content grew so quickly that in 2006 the U.S. publication *TIME Magazine* selected as their person of the year "You" to signify the rise in importance of the consumer voice online.

The incentives of user-generated content are both implicit and explicit. Implicit motivations are generally concerned with social relationships, and one example of this is Facebook friends. Another example of where a user implicitly seeks content is Yahoo Answers. The explicit motivations relate to tangible rewards that are usually financial in nature. This may include searching for new jobs or entrepreneurial advice.

The Organization of Economic and Cooperative Development (OECD) (2007) has identified three primary schools of user-generated content:

First, there is a publication requirement, and this may exclude e-mail and instant messaging.

Secondly, there is the creative effort put into the online publication. However, according to the OECD, this may be difficult to define.

Thirdly and finally, the work is created outside of professional routines and practices. This may relate to work being produced outside the normal time bands of work and professional duty.

Whilst these are general considerations, they may provide guidance for meeting and event planners when attempting to define for their guests the purposes and potentials of their generated comment.

During the past ten years there have been significant issues related to copyright and trademark infringement and bullying associated with user-generated content. Generally, in the U.S., copyright infringement is exempt with regard to user-generated content; however, in Europe the regulations are more complex. The issue of bullying, especially among adolescents, has grown in recent years with, sadly, even suicides being reported as a result of individuals receiving critical and negative comments online.

Therefore, while *vox populi* (the public voice) is critically important in terms of promoting a democratic state, it must be carefully cultivated, developed, managed, and evaluated to ensure that it promotes positive engagement among meeting and event guests in a sustainable manner.

Encourage and drive guest-generated content to your meeting and event website and through social media

Since the development of user-generated content in 2005 the invitation to "follow", "like", add a hashtag " become a friend", "post" has become commonplace at meetings and events. Literally millions of meeting and event guests have become further engaged in the content of their sponsoring organizations by connecting online.

One of the key issues regarding this engagement is the question of conversion. How does the meeting and event planner convert the guest-user content provider into an active attendee, volunteer, or leader? This evidence has not appeared to date; however, there does appear to be anecdotal evidence that the more engaged an association member is with their organization, the more likely they are to renew their membership. Therefore, the opportunity to encourage and drive guest-generated content to various social media platforms may ultimately lead to more committed members and participants.

According to Computer Hope (2014), when redirecting guests to your website, make certain you include a slash at the end of your URL, such as http://www.computerhope.com/. This will avoid re-directs to other sites and will send the guest directly to your website. Computer Hope further recommends personalizing your relationship with users by collecting birthdays and automatically sending them a Happy Birthday notice. Finally, Computer Hope recommends that you carefully consider the bandwidth, user, and the hardware they will use to view your content before you design your web pages or develop your content. Hardware and software may vary widely as will bandwidth, and therefore it is important that you customize your content to reach as many users as possible.

Bob Vaez, chief executive of Eventmobi.com, sees collaborative technologies as a major trend in the development of meeting and event technologies. Bob states that "We're already starting to see the personalized event experience take shape through event apps. One example is providing attendees with the ability to create a personal schedule via mobile app rather than circling sessions in a show guide. Another would be the ability to manage an event app personal profile which is essentially allowing attendees to manage their online presence at the event. The personalized experience will continue to improve as technology more deeply integrates with personal technology like smartphones and tablets" (2014).

He further argues that the growth in mobile technology has rapidly accelerated user-generated content. According to Vaez, "We're already seeing the impact mobile has on meeting planning with the ability to connect anytime, anywhere. This also allows planners to stay connected to their events and their attendees at all times, ensuring that everything is proceeding as planned. A quick text message from the other side of the conference hall or a tweet from disgruntled attendees allows planners and their teams to provide instant customer service. It's important to remember that positive experiences go a long way too; happy attendees will tweet their experience and share it with friends and colleagues."

Click Z (2014) has identified five effective methods for repurposing user-generated content:

1. **Interviews.** Raising questions is one of the most basic ways to crowdsource content. The method to employ depends on the desired outcome. Asking the community for suggestions of whom to interview and what questions to ask is a great way to get people involved. Interviewing industry thought leaders provides the brand's audience with unique content and creates a positive association between the "brandividual" and the company.

2. **Social Q&A.** Yahoo Answers, LinkedIn, and sites like Quora can provide very useful platforms to present questions and attract answers from a variety of people. Of course, your intent needs to be clear, and permission for reuse should be obtained before republishing. Those familiar with the Q&A communities can word questions to attract replies from specific individuals who might not otherwise respond to a content participation pitch via e-mail.

3. **Contests resulting in content.** Examples of contests in which consumers produce their own videos and share images abound on the social web. Search engines love any kind of content, especially text.

   Andy Beal runs a great contest for a search marketing scholarship on his Marketing Pilgrim blog (www.marketingpilgrim. com). The articles written by contestants drive traffic to Andy's website and also become content on it. To top it off, the articles are compiled into an e-book.

4. **Comment feedback loop.** One of the most meaningful ways for a community to engage with a brand is through comments on a company blog. Asking readers to participate in a dialogue by commenting can result in content that is better than the original blog post.

   Brands can then recognize blog commenters by drawing attention to the "best of" comments through a separate blog post or in a newsletter (at our agency, we do it in the TopRank Marketing Newsletter).

5. **Book authoring by community.** Reaching out to industry experts to share their insights as part of a larger project can be very effective. Author Michael Miller (2008) did this with Online Marketing Heroes, of which I was a part. He interviewed 25 successful marketers; the results of those interviews became the book.

   Another example involves soliciting subject matter experts to write articles of 1,000 words or so on predetermined topics. The brand serves as editor to compile the articles into an e-book, which can be used as fulfillment in a lead generation campaign.

   Meeting and event planners may use each of these five methods to engage and drive guests to their website and social media. You should fully exploit the democratization of the Internet by inviting your members or attendees to nominate individuals they would like to be interviewed or to vote for the top members in your organization. This provides an opportunity for your members and participants to organically develop your organization's brand identity.

   In Great Britain, there is a tradition of appointing a journalist to serve as an "Agony Aunt" and answer personal questions in the national newspapers. You may adapt this idea within your organization by inviting content experts and others to provide online advice for users.

You may further the brand identity of your organization by inviting users to create videos describing what they have learned at your conference or how they have benefitted from your exhibition.

Finally, you may collate all of the comments from your users and include them in an online book or article to provide a permanent record with potential analysis of their findings. Through each of these activities you are increasing the value of your organization or meeting and event for your participants. You are providing added value with minimal cost. Perhaps most importantly, you are leveraging the value of your existing content by introducing your conference, meetings, exhibition, and organization to many more potential stakeholders.

Moderate guest-generated content for your meeting and event to ensure appropriate comments are included

According to Cognizant (Khadilkar, 2015), 4.1 million minutes of video are uploaded to YouTube every day, and six billion images per month are uploaded to Facebook. However, 40% of images, and 80% of videos [created] are inappropriate for business. How do you carefully reduce the risk and moderate your user-generated content?

The three main methods for moderating your content include:

1. **Automated moderation**, using computer applications and algorithms.

2. **Community moderation**, leveraging the online community to self-moderate content (such as flagging or volunteer administration).

3. **Human moderation**, whether by a dedicated staff person or crowdsourced.

Automated moderation is very effective by identifying keywords and phrases that should not be allows to be included. However, as with any automated system, it must also be audited frequently to ensure that it is not ruling out the good with the bad content.

Community moderation is best exemplified by sites such as Wikipedia where users moderate the content to ensure that it is accurate and current.

Finally, human moderation, while the most time consuming and expensive, requires trained staff to view all posts and then approve them before they are allowed to be seen by others. This is most effective with a smaller online community.

According to Cognizant research (Khadilkar, Pai, and Ghadiali, 2014), human moderation can be extremely expensive. Figure 12.1 depicts the potential costs you may incur.

Many organizations have begun to outsource their content moderation tasks to reduce costs. Current research provides some evidence

Estimated Costs of Moderation

Content Example	Estimated Size	Estimated Moderation Time per item	Estimated Mechanical Automated Moderation Cost	Estimated Manual Moderation Cost
Video	6 minutes	1.7	$2.61	$277.00
Audio	6 minutes	1.4	$0.13	$230.00
Images	500 KB	0.4 seconds	$0.013	$0.70
Text	200 words	1 minute	$0.005	$167.00
Other? Video and audio mix, animation	Dependent on KB size	Dependent on KB size	Dependent on KB size	Dependent on KB size

Figure 12.1. Adapted by Goldblatt from Cognizant Research 2014.

that the earlier content is moderated (for example, using automated filters), the less likely there is to be a significant negative impact to the organization brand at a later date. Therefore, upstream monitoring of content results in less downstream problems later.

One of the best examples of the importance of content moderation is the website TripAdvisor.com. TripAdvisor.com was founded in 2000. It claims it is the world's largest travel site, with over 100 million users.

TripAdvisor has experienced many controversies in less than fifteen years. For example, 30 hotels have been blacklisted by the site for posting suspicious reviews, and one hotel in Cornwall, England, was accused of bribing guests to post positive reviews.

As recent as 2011, *The Guardian* newspaper (Sweney, 2011) reported that the United Kingdom Advertising Standards Authority stated that TripAdvisor must not claim that all of its reviews were from real travelers, are honest, and may therefore be trusted. The TripAdvisor example should serve as a loud and clear warning to organizations that promote user-generated content that strong disclaimers are needed to alert the readers that the opinions being offered do not necessarily reflect those of the host organization and that they may or may not be accurate and trustworthy.

In the U.S. hospitality industry there have been isolated examples of businesses being negatively impacted by harsh reviews on TripAdvisor and other sites. As a result, these same hospitality organizations often monitor these sites carefully and offer rebuttals as needed to provide readers with additional information to help protect their brand and reputation.

Meeting and event organizations should similarly monitor wherever possible those sites that that post reviews or other opinions regarding their programs and be prepared to counter these comments with a rational and objective description of the actual service that was provided. Further, it is helpful to have a meeting and event blogging team standing by to disseminate positive views to counter the negative ones whenever possible.

Promote democracy and transparency in terms of guest-related content

Although moderation is an essential part of the overall guest-generated content process, it is equally important to practice democracy and transparency. If you do not allow all sides of the argument, all views of the situation, and all feelings to be expressed, your readers may be suspicious that you are biased in your approach to their content.

Therefore, it is valuable to appoint a guest-generated content ombudsman who will represent the readers and investigate where necessary to ensure that all points of view are being allowed to be presented. Guests should be directed on your website to an e-mail address for the ombudsman, such as ombudsman@ABCassociation.org, to address their queries, complaints and suggestions. The ombudsman then should respond in a prompt manner following their investigation and provide constructive recommendations on how to remedy the issue to promote greater democratization and transparency in the future.

Avoid bullying and harmful comments through guest-related content

One of the major challenges of the twenty-first century Internet is the prevalence of online bullying, especially amongst adolescents. According to Bullying Statistics (2014), cyber bullying may include any of the following seven activities:

1. Sending mean messages or threats to a person's e-mail account or cell phone.

2. Spreading rumors online or through texts.

3. Posting hurtful or threatening messages on social networking sites or web pages.

4. Stealing a person's account information to break into their account and send damaging messages.

5. Pretending to be someone else online to hurt another person.

6. Taking unflattering pictures of a person and spreading them through cell phones or the Internet.

7. Sexting, or circulating sexually suggestive pictures or messages about a person.

This is a growing problem among young people as evidenced by the following statistics:

- Over half of adolescents and teens have been bullied online, and about the same number have engaged in cyber bullying.

- More than 1 in 3 young people have experienced cyber threats online.

- Over 25 percent of adolescents and teens have been bullied repeatedly through their cell phones or the Internet.

- Well over half of young people do not tell their parents when cyber bullying occurs.

According to the Cyber Bully Research Center, this problem has become more severe.

- Over 80 percent of teens use a cell phone regularly, making it the most popular form of technology and a common medium for cyber bullying.

- About half of young people have experienced some form of cyber bullying, and 10 to 20 percent experience it regularly.

- Mean, hurtful comments and spreading rumors are the most common type of cyber bullying.

- Girls are at least as likely as boys to be cyber bullies or their victims.

- Boys are more likely to be threatened by cyber bullies than girls.

- Cyber bullying affects all races.

- Cyber bullying victims are more likely to have low self-esteem and to consider suicide.

Therefore, meeting and event planners may be liable for the actions of others through the abuse of user-generated content to bully, blacklist, or damage the reputation of others. Organizations that offer user-generated content opportunities should encourage their users to immediately report any evidence of bullying to their ombudsman or web editor. The individuals who are being bullied should be told to save the negative messages they are receiving and then forward them to the person you have appointed to investigate within your organization.

Measure and evaluate guest-related content for evaluative purposes

Measurement is a continuous process through the meeting and event user-generated content journey. EventMobi's Bob Vaez states that "With the ability to track one's behavior and progress through a set of challenges or educational tracks, technology can help in the continuing learning experience after the event. Following up on the event experience has always been a challenge but with technology, it's possible to extend the experience with new, open channels of communication, constant updates of event material, and strategic promotions to tie events together year-after-year" (Vaez, 2014).

There are numerous ways to measure and evaluate guest-related content within your organization. However, most corporations and associations find it very difficult to measure. Figure 12.2 depicts where they focus their evaluative efforts in order to obtain a return on objective, according to a recent study of corporations that evaluate their social media presence.

The use of Facebook likes and clickthroughs and retweets dominates all other metrics. This organization estimates that the cost of the transaction for social media is equivalent to five percent per sale.

Therefore, while the scientific methodology for measuring the value of guest-generated comment through social media is continually developing in the twenty-first century, it is important for your orga-

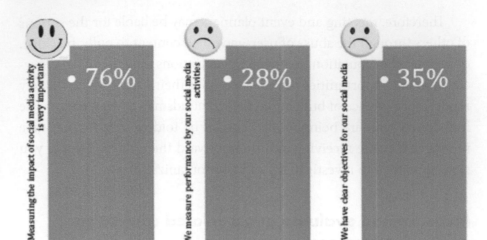

Figure 12.2. Focusing upon Evaluation Efforts (Adapted by Goldblatt from marketingcharts.com 2012.)

Measurements Used by Brands to Evaluate Effectiveness of Social Media Content According to United States Marketers of Corporate Clients

Likes 89%

Clickthroughs 87%

Retweets 81%

Daily or monthly active users 66%

Reach 66%

Conversation volume 52%

Influence 41%

Advocacy 27%

Sales 24%

Return on Investment (ROI) 23%

Cost per conversation 18%

Cost per sale 5%

Figure 12.3. Performance of Social Media by Brands (Source: emarketer.com 2014).

nization to seek your own customized return on objective for each activity you design and implement.

For example, you may seek to attract more nonmembers to your social media and guests user content sites. Further, you may wish to convert a greater percentage of casual visitors to become members or attendees at your conference. Therefore, it is important that you identify specific objectives for each task and then measure your performance in a real time manner.

SUMMARY AND CONCLUSION

In less than ten years, the growth of guest-generated content has become one of the most important trends in the twenty-first century meetings and events industry. Through the widespread availability of mobile phones and Wi-Fi, the average meeting and event attendee is able to multitask anywhere and everywhere. Part of this multitasking relationship should include a continuous conversation with your organization to deepen their relationship with your content. By providing them with the invitation, accessibility, tools, safety protocols, and moderation, you may not only rapidly engage and convert more participants but also greatly improve your overall brand identity.

DISCUSSION QUESTIONS

1. How will you use a content expert to answer questions about your meeting and event content in a real-time manner during your program?

2. How will you moderate comments to promote democracy but protect and preserve your meeting and event brand image?

TASK

Create a guest-user quiz about your meeting and event content to engage your participants before, during, and after the meeting. Using Figure 12.3 as a guide, create an evaluation metric to increase the number of likes, clickthroughs, and retweets from your meeting and event.

REFERENCES

Berners-Lee, Tim, Fischetti, M. (1999) *Weaving the Web: The Original Design and Ultimate Destiny of the World Wide Web by Its Inventor.* Retain: Onion Business.

Bullying Statistics (2014). Viewed: http://www.bullyingstatistics.org/content/cyber-bullying-statistics.html

ClickZ (2014). Crowdsourcing and user-generated content. Viewed: http://www.clickz.com/clickz/column/2098809/crowdsourcing-user-generated-content

Khadilkar, Abhijeet, Pai, Tom, and Ghadiali, Shabbir (2014). *How to De-risk the Creation and Moderation of User-generated Content. Cognizant.com.* Viewed: http://www.cognizant.com/InsightsWhitepapers/How-to-De-Risk-the-Creation-and-Moderation-of-User-Generated-Content.pdf

Computer Hope (2014). Viewed: http://www.computerhope.com/promote.htm

eMarketer.com (2014). *Measures Used by Brands to Promote the Effectiveness of Their Social Media Content by Client Side Marketers in April 2014.* Viewed: https://www.emarketer.com/Newsletter

Levine, S., Prietula, M.J. (2013). *Open Collaboration: Principles and Performance. Organization Science.* Reviewed: http://doi.org/rfb

MarketingCharts (2012). Viewed: http://www.marketingcharts.com/direct/most-companies-finding-social-medias-impact-tough-to-measure-23213/econsultancy-measuring-social-media-impact-sept-2012png/

Miller, M. (2008) On Line Marketing Heroes New York, NY: John Wiley & Sons, Inc. Nazarian, V. (2010) *The Perpetual Calendar of Inspiration.* Spirit.

O'Connor, P. (2008) *User-generated Content and Travel: A Case Study on Trip Advisor.com.* Information and Technologies in Tourism.

Organization of Economic and Cooperative Development. (2007) Viewed: http://www.oecd.org/dataoecd/57/14/38393115.pdf

Sweney, Mark (2011). *ASA To Investigate TripAdvisor.* London: Guardian. co.uk.

Vaez, B. (2014) Eventmobi. Personal communication. Toronto, Canada: Eventmobi.com.

CHAPTER 13

Registration and Transaction Systems

> "If you wrote something for which someone sent you a cheque, if you cashed the cheque and it didn't bounce, and if you then paid the light bill with the money, I consider you talented."
>
> —Stephen King

LEARNING OUTCOMES

As a result of reading this chapter, you will learn how to:

- Utilize advantages of online event registration system

- Select secured event registration system

- Evaluate existing payment systems and identify key components of the payment system

INTRODUCTION

How technologies change meeting/event registration

The history of meeting and event management registration systems is one that spans barely fifteen years and yet is fraught with explosive growth. According to Rick Borry (2014), one of the first online meeting and event registration systems was titled Enteronline.com and first appeared in 1997. This system was used to register runners for sports events. Borry then launched Register123.com in 1998 to create an online registration platform for business events.

By the end of 1998 Regweb had a working registration system and was soon followed by B-There.com. These platforms were rapidly followed by Starcite, Plansoft, SeeUThere, Evite, Acteva, Regonline, and ultimately Cvent.com. According to Borry, although the rapid development of new products created a cluttered marketplace, few organizations really understood the potential for the online registration system. However, in the latter part of the twentieth century George Little Management, one of the largest producers of trade shows in the United States, adopted Register123 as their online registration system, and they continue to use online registration fifteen years later.

Borry states that one of the most important considerations in the development of online registration systems was trust. The major issue was the replacement of trust in a physical site to one that is virtual. However, following the economic decline in the meetings and events industry of 2001, there was a need to automate registration procedures due to large numbers of reduction in staff in the meetings and events industry. As a result of this need and the growth of online transactions, trust naturally developed among meeting and event registrants who sought faster, easier, and more efficient ways to register and complete their transactions.

GROWTH OF ONLINE MEETING AND EVENT REGISTRATION

In 2000, Professor Joe Goldblatt predicted at the "Beyond 2000" conference for event educators in Sydney, Australia, that the majority of registrations would shift to online within the first decade of the twenty-first century. Due to the various registration products identified by

Borry, the need by meeting and event organizations to reduce costs and the desire by meeting and event attendees to save time and increase efficiency, this prediction has come true.

There are still challenges with online registration systems, including security, encryption, and ease of download due to limited broadband capacity of users. However, it is predicted by many meeting and event planners that online meeting and event registration will continue to grow rapidly throughout the twenty-first century. One of the major proponents of this growth has been the explosion of mobile technologies throughout the world. The opportunity to register for a meeting or event using 3 or 4 G any time and anywhere is very attractive to busy millennials who often combine work and leisure experiences. Therefore, it is essential that meeting and event planners understand and master the potential usages of online registration systems to best benefit their attendees and their organizations in the near and mid-term future.

ONLINE REGISTRATION SYSTEM

As Internet use becomes increasingly common, websites have become essential attributes for events companies, which resulted in an increased use of online registration. Online registration software can maximize event administrative staff productivity and increase the number of attendees from improved response rates. Handling registrations by paper is now rarely done; it can cause inaccuracy of data; confusion of details; the need to manually edit registration information, cancel a registration and issue a refund; and inefficiency in processing and confirming that many event organizers had to suffer. Fortunately, online payment along with online registration have developed quickly and easily. Today's events are promoted via online-based and branded event websites, and attendees register an event via linked online registration pages. To establish a secure and easy-to-use online registration system, there are necessary components of an online registration system.

COMPONENTS OF ONLINE REGISTRATION SYSTEM

Online registration system

There are two ways to create an online registration system. One is to develop your own registration system. This is ideal for a case of an ongoing (reoccurring) large annual event/meeting. This is particularly beneficial as it is easier to develop a tailored registration page per needs. While there is an initial development cost of an organization's own online registration system, it will meet the break-even point eventually, and after that there will be virtually no additional cost for the system (except some maintenance or ongoing upgrade, if necessary).

The other option is to subscribe to an online registration service operated by third-party suppliers. Third-party online registration systems (further discussed later in this chapter) charge fees per number of registrants. Therefore, a meeting/event organization that hosts an annual or recurring event with a large number of attendees will end up paying a large amount of online registration service fees year after year. This option is, however, ideal for a small organization that doesn't have many resources (e.g. IT professionals and infrastructure) and which hosts occasional and nonrecurring events and meetings. As the cost of online registration service will be by the number of registrants, and security and technical support is all included in the charge, this may be the most economical and secure online registration solution for these organizations.

While not every event needs a custom design, it is a great way to set registration pages and event pages apart from others and to create a seamless experience from the organization's website to the event. Details of third-party online registration service providers will be discussed in later in this chapter.

Printing name badges and labels

Once registration is closed and the necessary information has been collected from attendees, the next step is to create custom name badges and mailing labels. This step is now integrated with online registration system so name badges and mailing labels can be quickly and easily extracted from the attendees' data to create name badges for attendees.

Security protocol

The importance of securing attendees' information has never been more important than now. We hear news constantly of customers' confidential information (including financial information) being breached by hackers. This results in millions of dollars lost to the company and loss of credibility from their customers. The same level of attention to data security should be paid to the event registration system. Advancements in online payment processing are more secure than ever and provide peace of mind to both attendees and event organizers. The Payment Card Industry (PCI) (an information security standard for organizations that handle branded credit cards from the major card schemes) regulates online payment security and procedure in various transactions. PCI Compliance Level 1 is one of the highest level of security, and it is suggested by many online registration system.

Additional functions

Today's meeting and events offer various activities outside of formal programs. They include special event tickets, day trips, and pre- and post-tours. It is important that an online registration system provides easy navigation through all of the event's offerings. The online registration should be designed to accurately track and confirm an individual event as an attendee sign up for different sessions or break-outs. In addition, online registration can lead to access housing sites and travel (airline, rental car, train, etc.) solutions for easy management of event travel planning.

ADVANTAGES OF ONLINE REGISTRATION SYSTEM

Save time and accuracy of registration data

Instead of collecting information from attendees and keying that data into another program, online registration system lets attendees enter their own information. This helps to avoid mistakes by having attendees enter their own information directly into the system so that information is not misinterpreted or misplaced by event organizers on the other end.

Imagine that for today's global events and meeting, attendees' address formats, names, and job titles may be all different by the origin of attendees. For example, one of the author's friends, who is from Thailand, has this last name:

NI-RAT-PAT-TA-NA-SAI

It is almost impossible to enter this last name without a typological error by a registration staff in U.S. Similar stories apply to addresses. Another example: the address of South Korea is something like this:

456-3 Kaeyang-dong, Pupyong-ku, Inchon metropolitan city, South Korea 403-113

An online registration form can allow attendees from around the world to enter their accurate information by setting up structures for entering information field specific to an attendee's country of origin.

Easy change/cancellation of registration

Online registration systems offer services not only to register but also to modify, cancel, and refund attendees. Changes are easily made and communicated to attendees. Some systems even allow registrants to do these operations themselves. This can save a tremendous amount of time and labor by the use of a confirmation number or e-mail address.

Multi-track registration

Online registration systems can build a dynamic, multi-track, multi-session event registration process. They can provide options to let attendees sign up for different "tracks" or themed sessions, or even choose from different sessions to make their own dynamic track throughout the course of the overall event.

Easy duplication of a past event registration page

Many events are reoccurring ones. An online registration system allows cloning previously created events easily. Past event registration pages can easily be duplicated and updated for a new event. The ability to copy event links from one event to the next can save a significant amount of time and minimizes redundancies across similar events.

Easy filtering of registration

Online registration systems can be used to restrict registration by filtering the number of registrants, e-mail addresses, registration codes, or unique IDs assigned to each invitee. There are meetings and events that offer some information or particular sessions to only assigned group of invitees or to only a limited number. An automated wait list can be created to limit capacity to the event or to particular sessions. With the options made available by an online registration system, setting these restrictions and permissions can be easy.

Instant confirmation

Online registration can provide an instant confirmation of registration to attendees via e-mail and a registration page along with a registration receipt (if applicable). Processing payments is simple, and all data is easily accessible. This prompt process can give attendees an assurance of their successful registration and the processing of their payment.

Attendee tracking and customized report

With an online registration process, customized reports can be pulled in an instant. Event and meeting planners can easily access detailed reports leading up to events to better understand attendee segments. This can help event/meeting planners customize events for their audience.

Survey

The survey tool can be integrated within an online registration system to track attendee satisfaction and allow attendee suggestions in order to improve programs year after year.

ONLINE TRANSACTION/PAYMENT SYSTEM

Among many event operation tasks, handling processing payments and refunds manually can be one of the most challenging tasks. An online transaction/payment system provides collecting payments quickly and transmitting them securely into an event organizer's bank account. It should allow registrants to choose alternate payment options, such as payment by check, purchase order, or wire transfer. The online transaction/payment system also needs to allow purchases of other products and services for its attendees. Today's events usually offer various extra fee activities and merchandise (e.g. event T-shirts, books). Therefore an online transaction system should be able to integrate with an existing merchant account and virtually any e-commerce gateway. It is important to confirm that an online registration company doesn't charge a transaction fee as a middle man and that event registration payments and other transactions are directed into an event organizer's financial account.

Security protocol

The most widely adopted payment policy is the Payment Card Industry Data Security Standard (PCI DSS), a set of requirements designed to ensure that companies that process, store, or transmit credit card information maintain a secure environment. Essentially any merchant who is certified as PCI Level 1 compliant, the highest security rating issued by the PCI, can provide assurance of security to users of online transaction systems, and event organizers can have confidence in data security.

SCREEN SHOT

The Payment Card Industry Security Standards Council (PCI SSC) denotes the debit, credit, prepaid, e-purse, ATM, and POS cards and associated businesses. It is is an open global forum for the ongoing development, enhancement, storage, dissemination, and implementation of security standards for account data protection. The council was originally formed by American Express, Discover Financial Services, JCB, MasterCard and VISA on Sept. 7, 2006, with the goal of managing the ongoing evolution of the Payment Card Industry Data Security Standard.

The PCI Data Security Standard is comprised of 12 general requirements designed to build and maintain a secure network; protect cardholder data; ensure the maintenance of vulnerability management programs; implement strong access control measures; regularly monitor and test networks; and ensure the maintenance of information security policies.

A Payment Card Industry Data Security Standard (PCI-DSS) Level 1 is the highest possible security level in accordance with the PCI-DSS. More information is available here: www.pcisecuritystandards.org.

ADVANTAGES OF ONLINE TRANSACTION/PAYMENT SYSTEM

■ Logistical headaches often stem from manual payment transactions. Instead, collecting online payments can provide following advantages:

Easy setup of various registration fees

Events and meetings use variable pricing (e.g. early registration discounts, group registration discounts, member vs. non-member fees), and this can be a challenging task when using a manual transaction system. However, variable pricing can be easily set up using an online transaction/payment system by using dynamic pricing based on registrant type and key due dates.

Faster receivables

Online transaction/payment processing makes collecting event fees convenient for registrants and planners alike. Online transaction/payment systems allow faster deposit of registration fees and other dues by attendees for meeting/event organizers, with real-time tracking and automated refunds. It is very critical as many payables of event/meeting planning are due before the actual event, so cash flow is incredibly important to operating event expenses. In addition, deposited funds can generate interest, which can be an extra revenue item. For example, 1,000 attendees paid early-bird registration fees at US $1,000 each, which is total of $1,000,000 in a savings account. If the savings account has a 1.9% interest rate, an event organizer can earn $156 in interest earnings each month. Once an attendee clicks the "confirm payment" button from the registration page, the fees will be immediately deposited into the event organizer's account.

Processing various forms of payment

As meetings and events are a global business, there are many attendees and suppliers from around world. That means that international currencies will be used, and an event organizer should be prepared to handle those. In addition there will be various formats of payment such as major credit cards as well as payment by check, cashier's check, purchase order, or wire transfer.

Many of the aforementioned forms of payment must be deposited to the physical bank by delivering them physically, and it takes more time

  CyberSource Authorize.Net

Figure 13.1. Online Transaction E-commerce Gateways

for those funds to become available (some may take weeks to be cleared if the payment is issued by an international financial institution).

Integrating with e-commerce gateways

Online transaction/payment system providers can connect an event organization's existing merchant account or acquire a new merchant account for seamless payment processing. They provide a quick payments process via e-commerce. The most utilized online payment service providers include Verisign, Cybersource, Touchnet, or PayPal Express.

Quick and automated refunds

Attendees are impatient especially when it is about refunds. Refunding money quickly and accurately can be done effortlessly by an automated refund procedure in the transaction/payment system. It can help to minimize the workload by establishing an automated refund structure that is contingent upon the date someone cancels a registration.

LEADING ONLINE EVENT AND MEETING REGISTRATION SERVICE PROVIDERS

CVENT (www.cvent.com)

First established in 1999, Cvent is currently the world's largest online event registration service, with customers in approximately 100 countries and 1,450+ employees worldwide. Cvent online event registration software can increase attendance to events and decrease your costs. Cvent's registration

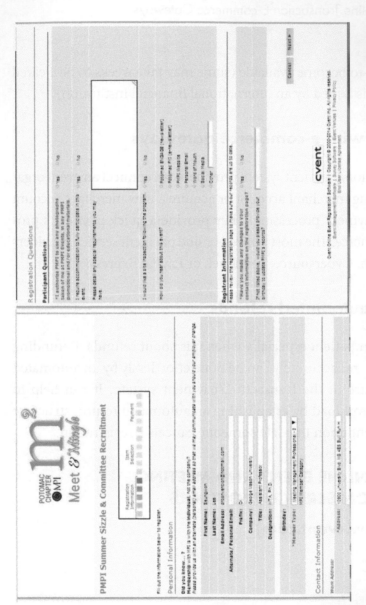

Figure 13.2. A Screen Shot of a Cvent Online Registration Page (Courtesy of MPI and CVENT.COM).

solution combines advanced functionality with a user-friendly interface, allowing you to design a specialized registration process for your audience.

Features

Cvent provides a large selection of attractive predesigned registration page templates. Meeting planners can use these pages as is or use them as a starting point for customization. It provides custom name badges and mailing labels for a professional look with predesigned name badge layouts. It allows uploading exhibitor rules, brochures, or trade show layout maps. Attendees can submit registrations via web and mobile devices. Planners can manage attendee communication by integrating the event registration data into the Cvent event management system. It also can produce custom registration status reports with pre-populated forms. Its payment module is certified by PCI Level 1 compliance.

ACTIVE RegOnline (www. activeevents.com/ solutions/product/active-regonline)

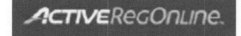

ACTIVE RegOnline is the entry-level, online event registration software for event planners of all levels and events of all sizes. The web-based application requires no set-up costs and no software to install, and no contracts are required. This system offers performance-boosting tools for easy event promotion, attendee management, data tracking, and event management websites, plus hotel and venue sourcing tools. Technical support is provided from an all-in-one central location.

Features

Flexible online registration forms can streamline even the most complex event registration processes, allowing attendees to easily apply discount codes, sign up for sessions, and access lodging and travel information.

Attendee management: Easily access, update, and share important event information with real-time access from a PC or mobile device. Registrants can also log in at any time to update their online registration details and access event information.

Payment processing: Securely manage a wide range of payment types with various currency options, discount codes, and event registration types. Save time with automated purchase confirmation e-mails and receipt processing. It is certified by PCI Level 1 compliance.

Onsite event tools: Ensure that your attendees have a positive experience from start to finish with tools for onsite event badge printing as well as self-service kiosks for automated onsite check-in and registration.

EVENTBRITE (www. eventbrite.com)

Eventbrite is a free online event registration service that provides all the tools an event/meeting planner needs to make an event registration page. It is very simple solution to organize events/meetings.

Features

Eventbrite offers predesigned templates with different looks for an event page. It also allows an event/meeting organization to offer multiple ticket types, discount codes, and merchandise. It can process credit cards and PayPal for online registration payment as well as track payments and manage refunds. Its convenient unique feature is an ability to handle onsite check-in and ticketing using mobile apps. It can also produce an updated attendees list for check-in using its free Entry Manager App and sell tickets onsite using its mobile box office.

SECURITY OF ONLINE PAYMENT AND ATTENDEE INFORMATION

As the majority of meeting/event registrations have moved to online, collection of secure online payments has become very critical in meeting management.

Secure Online Payments

Setting up secure, automated payments and refunds is very critical in online registration management. This requires secure processing of fees via Payment Card Industry (PCI) compliant tools. As mobile-based payment is growing, it is of increasing importance to provide secure collection of registration payments and other important transactions via secure mobile.

Security of Attendee Information

A security policy for handling attendee information must be in place to prevent any improper access or damage of attendee information. Proper handling is sometimes mandated by law. Legal and professional requirements apply to sensitive personal information (e.g. medical information, financial information) about attendees.

The good news for event/meeting organizations is that most online registration systems are now run by third-party online registration providers using a cloud computing system, which provides better management of data security and loss prevention. It is critical to maintain a high level of secrecy as it is concerned with improper disclosure of information.

NEW TREND: MOBILE REGISTRATION SYSTEM/APPS

Smart mobile devices (e.g. iPhones, iPads, Android-based smarthones, Google phones, etc.) are being used in many meeting/event management functions. Therefore, many online registration service providers offer mobile-based registration and attendee management services. They include:

- **Mobile event apps:** A branded mobile event app enhances attendees' registration experience. The attendees can manage their event registration, access up-to-date event information, and improve networking opportunities via a mobile connection.

- **Mobile attendee management:** An event/meeting planner can access all attendee and event information from his/her mobile device and make quick updates wherever and whenever. It can be utilized for on-site registration and session check-ins.

CONCLUSION

The rapid growth of online registration systems for meetings and events is well documented. What is not so well understood is how these systems will be protected and regulated by governments as the issue of individual privacy becomes of greater and greater concern to politicians, bureaucrats, and users of these systems.

In this chapter you have explored the brief but explosive history of online meeting and event registration systems. You have also learned the purpose of these systems and their potential for increasing bottom-line revenue for your meeting and event organization. Finally, you have learned why security and privacy are increasingly important and how you may adopt and develop procedures to help reduce your risk through information gathering.

DISCUSSION

Discuss with others how you would consult with your potential meeting and event participants to identify their needs, wants, and desires before developing an online registration system. What types of questions will you ask them? Then determine, based upon their responses, whether you will use an existing online registration system or if you will create your own customized system for your meetings and events.

TASK

1. Work with a small group and use Cvent.com or Eventbrite.com to create a test online registration platform for your event. Once you have developed the site, ask the users for their feedback.

2. Discuss importance of security (payment, attendee information) and role of PCI.

REFERENCES

Borry, R. (2014) *The Registration Doctor.* Viewed: 2014. Accessed: http://registrationdoctor.blogspot.co.uk/2007/08/history-of-online-registration.html

Goldblatt, J. (2000) Beyond 2000. Events Beyond 2000. Setting the Agenda. Sydney, Australia: University of Technology Sydney.

ADDITIONAL RESOURCES

www. activeevents.com/solutions/product/active-regonline – Active RegOnline

Cvent. www.cvent.com

Payment Card Industry. Accessed Security Standards Council www.pcisecuritystandards.org

CHAPTER 14

Crowdsourcing for Events

> "There are reasons to believe that the current manifestations of crowdsourcing are just a prelude to a far more pervasive transformation. Actually, there are about 200 million reasons to believe it. That's the rough number of kids around the world that currently have Internet access".
>
> —Jeff Howe, a contributing editor at Wired magazine (Howe, 2009)

LEARNING OUTCOMES

As a result of reading this chapter, you will learn how to:

- Understand the power of crowdsourcing
- Use crowdsourcing for your events
- Plan a crowdsourcing strategy
- Evaluate your crowdsourcing efforts
- Find the best tools for crowdsourcing
- Understand the principles of crowdfunding
- Use crowdfunding for your events

INTRODUCTION

The environment in which we live today provokes us not only to be part of an audience, but also to step on the stage and participate in what's happing there. New technologies provide opportunities for members of the audience to share comments, then express an opinion, and nowadays, to take an active part in the decision-making process. Without any doubt, the Internet gives people around the world a medium where they can easily collaborate with one another. Witnessing this process, businesses started looking for ways to take advantage of this growing power. In other words, many have started considering how to use the creative and productive capabilities of Internet users to obtain their tangible objectives. In the process of working out the best possible way to get hold of the power of users, many companies have started asking questions; some have even asked for help. It was difficult in the beginning, as companies had to admit that someone outside their company could come up with a solution, a proposition, or an idea that turned out to be better than the ones suggested by their own staff. However, following the steps of other successful businesses, companies have realized that the change is ongoing and is here to stay. All this resulted in their establishing close relationships with audiences, facilitating the participation of consumers in the making process. The latter model was used by private companies, NGOs, governments, and different organizations.

The contributing editor of *Wired Magazine,* Jeff Howe, along with Mark Robinson coined the term "crowdsourcing" in June 2006. In an article published in *Wired,* Howe asks the following question: Remember outsourcing? Sending jobs to India and China in 2003? The new pool of cheap labor is now everyday people using their spare time to create content, solve problems, even do corporate R&D. That phenomenon is called crowdsourcing. According to Howe, "Simply defined, crowdsourcing represents the act of a company or institution taking a function once performed by employees and outsourcing it to an undefined (and generally large) network of people in the form of an open call... The critical prerequisite is the use of the open call format and the large network of potential laborers" (Howe, 2006).

Crowdsourcing would not be so popular and easy to use without the Internet. In fact, it is the Internet that allows us to communicate fast and cheaply with a large network of people. According to Daren C. Brabham, "Crowdsourcing is not merely a web 2.0 buzzword, but instead a strategic model to attract an interested, motivated crowd of individuals capable of providing solutions superior in quality and quantity to those that even traditional forms of business can" (Brabham, 2008).

It is important to point out that the process is sponsored by an organization, and that the working process of the crowd is directed by this organization. Crowdsourcing can be defined as a creative top-down managed process. In addition to this, in most cases the organization makes a considerable profit from the work of the crowd and, consequently, offers different rewards, even individual prizes, to its most efficient and conscientious workers.

Due to the fact that crowdsourcing is at the top of some business agendas, many specialists study the phenomenon to find out which factors motivate people to be part of that new crowd. If we look at the top ten reasons and try to incorporate them into the event industry, we will find different ways to use crowdsourcing in events as well as what precisely motivates the crowd to participate (Table 14.1).

Depending on the purpose of crowdsourcing, in general, we can define four different types.

Type 1: Crowd Creation. We ask individuals in the crowd to create a video, translate from one language to another, solve challenging scientific problems, etc. Crowdsourcing can be effective not only for sourcing new writing, photography, music and film, ideas

Table 14.1. Motivations for Individuals in Crowds and in Events

Business	Events
Earning money	Earning free pass for an event
Develop creative skills	Develop new skills
Networking with creative professionals	Networking with potential partners or employers
Build personal portfolio	Build professional portfolio
Personal challenge to solve problems	Practicing problem-solving process
Socialize and make friends	Develop professional relationships
Passing time	Professional development in free time
Contribution to a large project of interest	Adding value to an event of common interest
Sharing with others	Sharing knowledge and expertise with others
Having fun	Having fun

for products, speakers and places for events, but also for solving real-world scientific problems. For example, one of the problems for which crowdsourcing can find a solution to is the need for fresh ideas. As specialization in any field is a must nowadays, sometimes a lack of experience is the key ingredient for new ideas. Take, for instance, the success of the song "Happy" by Pharrell Williams and the video release by the name of "Pharrell Williams – Happy – We Are from [name of the city]". As of May 2014, more than 1,950 videos have been created. Inspired by this global phenomenon, a French couple launched a website wearehappyfrom.com to showcase the remakes. That's what the crowd can create for your event. Of course, you have to hit upon an inspiring idea that will motivate the crowd.

Type 2: Crowd Voting. In that way we can leverage the community's judgment to organize, filter, and stack-rank content. Crowd voting is the most popular form of crowdsourcing, which generates the highest levels of participation. We can apply the 1:10:89 rule, which states that out of 100 people:

- 1% will create something valuable

- 10% will vote and rate submissions

- 89% will consume creation.

For 10% who vote and rate content, "the act of consumption was itself an act of creation." Think about the countless opportunities to engage potential guests in voting activity in the early stages of the planning process of the event. For example, the crowd could vote on the topic of the event, the keynote speaker, the place, the duration, and many other different options, including voting for the best picture, best cover, and best talk.

Type 3: Crowd Wisdom. This refers to harnessing the knowledge of a considerable number of people in order to solve problems or predict future outcomes or help direct an event strategy. Studies confirm that crowds consistently outperform even concentrated groups of highly intelligent people. According to Jeff Howe (2006), "Given the right set of conditions, the crowd will almost always outperform any number of employees—a fact that many companies are increasingly attempting to exploit." We will not go into detail in discussing the reasons for this, yet it is important to keep in mind that every time you are facing a difficult situation, the crowd can offer useful advice and good solutions.

Type 4: Crowd Funding. Provided that the traditional corporate establishment is not an option for funding, nowadays you can use other people's financial support to execute different ideas. In the event industry this has been used for a long time, as we can use early-bird offers to fund our event in advance. There is a wide range of platforms used today that help events to find funding. They will be discussed in greater detail later.

Using the power of the crowd is not an idea we have suddenly come across. In fact, we have always asked friends for help. Now, thanks to networks on the Internet, there are many more real or virtual friends and followers whom we can turn to for help.

Taking advantage of the huge potential and unlimited capabilities of the Internet, we can make common cause with people from all over the world who can help us develop our projects and vice versa—do the same for others. What's more, now we can benefit from the power of the most capable and intelligent people on earth, not only on those in the office, as long as we know exactly what we want and how to motivate individuals. People, who were once known as being merely an audience, behave as participants now. They are ready to share their knowledge, time, and efforts if they believe in the benefits.

What we need to be aware of when we start planning a crowd-sourcing strategy is to take into account some basic principles of the crowd. First of all, the crowd is everywhere, and while the crowd can easily be brought together online, it is very difficult to do the same offline. Second, people in the crowd are multitask-oriented, which means that we have to suggest such a task that will require less than half an hour of their attention. Third, as we count on all people, we can find a lot of experts willing to help us. Next, we will not be able to use most of the results. Nevertheless, that same crowd is very good at finding the best solution. What does this mean for event practitioners? At first we can receive ideas from all over the world, provided that we ask for the right task and offer the right rewards. We have to divide a tough task into smaller ones, which people will be able to solve in less than 30 minutes. Although smart people, professionals with considerable expertise, are part of the crowd for a number of different reasons, we should not underestimate the level of specialization of such individuals in the crowd. On receiving solutions, we should expect that a lot of them will be unusable and valueless, so establishing a mechanism for the crowd to evaluate the best ones is of great importance.

Now, let's turn to the ways we can use crowdsourcing in the event leadership process. According to Dr. Joe Goldblatt, all successful events have five critical stages in common to ensure their consistent effectiveness.

Figure 14.1. Event Leadership Process (Source: Goldblatt, 2014)

These five phases or steps of successful event leadership are, respectively, *Research, Design, Planning, Coordination,* and *Evaluation* (see Figure 14.1)

At first glance, we can easily decide that crowdsourcing is not applicable to all five phases, but after more careful considerations matters might look different.

Phase One: Research. Generally speaking, we conduct research prior to the event because our goal is to produce an event that matches best with the planned outcomes of organizers or stakeholders. In the course of the planning process, we have to collect different data about our potential guests, such as gender, age, and income. We conduct a written survey to find out different attitudes and opinions. Moreover, we have to examine the culture of the community in order to identify comparable characteristics as well as to collect demographic and psychographic data. Thanks to the Internet we can do that much more quickly and at a very reasonable price. Just imagine what a dedicated and well-motivated crowd can do for us at that phase. In fact, with the help of the crowd, we can collect all the data on condition that we ask the right person. What's more, we have to ask for the right task and offer the right reward. We can start first by asking participants from the last event and expand the database by including people they recommend. Thus, with their help we will be able to count on a much greater number of participants. Besides, by using various online tools for conducting interviews, pools, and conversations, thereby delegating the crowd to do the task for us, we are in a position to collect better data than ever before. We can even rely on the crowd for the interpretation of those results. Let's take, for example, the principle on which platforms like Coursera work. Coursera is a for-profit educational technology company offering extensive open-online courses founded by computer science professors Andrew Ng and Daphne Koller, from Stanford University. Due to the huge number of students, the professors are not able to read all papers, so they delegate the task to students, giving them strict rules to follow. We can apply the same principle to the crowd to summarize the results or other kinds of data. It is obvious that by using online technologies and the help of the crowd, the cost of our research will be less expensive yet with better quality.

Phase Two: Design. Most people decide to start a career in the field of events because of the creativity involved in organizing them. Perhaps the best way to enhance your creativity continually is to surround yourself with highly creative people. Whether you are in a position to hire creative individuals or must seek for creative types in groups outside the office, you must find the innovators in order to be innovative yourself in your field. It is crowdsourcing that gives you a unique opportunity to tap not only into the creativity of your team but also to involve the participation of people from all over the world through the timely application of discussions and focus groups and, at the same time, relying on all other sources of information available to you thanks to the Internet. A reasonable event practitioner would avail of all the benefits of modern technologies in order to take advantage of the knowledge and expertise of the finest professionals in their respective fields. Last but not least, for questions of great significance, which you have to analyze at this phase—namely, what, why, who, when, and where—you can use the power of the crowd. In most cases event professionals ask former participants various questions. However, nowadays we can easily turn to future participants and professionals. Mind that the question "why" we should do an event is a crucial one. After finding its answer, we can use the help of the Internet society to meet more adequately the needs and expectations of guests who will be attending our event. Thus, we can meet the questions of who, when, and where more confidently and with greater precision.

At the same time we have to create criteria for the evaluation of all ideas that we will come across, as we are well aware of the goals of our event and what results to expect in the end.

- **What** refers to questions about ideas related to the format of the event and the adjustment of our plans.

- **Who** refers to questions about ideas for speakers, musicians, hosts, etc., or what kind of participants we need in order to reach new potential guests.

- **When** is an easy question considering its numerous options— at least 365 per year. We have to use this query to find out the motives of our respondents, which would help us the most appropriate date.

- **Where** is an important question because even an experienced event specialist might neglect a very popular place that their target group would rather choose to frequent.

All in all, the second phase is the one that can mostly benefit from crowdsourcing.

Phase Three: Planning. As we are well aware, the planning phase involves using time/space/tempo laws. That phase is closely related to the work of your team, and crowdsourcing can be used to find the best solution concerning more challenging tasks. Reliance on your own team is particularly important at this phase. A good point to consider is how to use crowdsourcing for the promotion of your event. Such tactics have to be planned carefully and executed with great accuracy.

One of the most conspicuous examples of how to use the power of the crowd can be derived from Barack Obama's campaign for President in 2008. Barack Obama has been called *Advertising Age's* marketer of the year for 2008 (Creamer, 2008). Mr. Obama won the vote of hundreds of marketers, agency heads, and marketing-service vendors gathered at the Association of National Advertisers' annual conference. One of its strongest and most impressive features of the campaign was the way it motivated the crowd to work for the sake of the campaign. In 2009 PR agency Edelman published "Barack Obama's Social Media Toolkit", where you can find a detailed analysis of the strengths and weaknesses of election campaigns (Lutz, 2009).

Phase Four: Coordination. This is the most obvious part of event planning, which is the real event itself. How can crowdsourcing help us at this point? If you take into consideration the enormous power of volunteers, you will find the answer you need. Human resources should be at the focus of our attention because they are always in demand, being crucial for the success of any event. The power of the crowd can help you discover people with a strong sense of dedication who would be willing to help you in conducting your event. So, do not hesitate to ask for volunteers because there are always people who will be motivated enough to become part of your event. What you need to do is find them and offer the best proposal and suitable rewards. It looks quite easy, doesn't it? On many occasions volunteers make up an integral part of the event

Phase Five: Evaluation. It is not so easy to leave the last phase of the process to the power of the crowd. That's a task that you have to do by yourself. Of course, there is a way to get help. For example, participants can be encouraged to make and share photos and videos, which can be used in the evaluation process. Modern event evaluation techniques allow us to use different methods for evaluation, while pictures and videos can help us to detect the participant's point of view.

To sum up, although crowdsourcing can help us in every phase of the management process, it is most efficient during the research and design phases.

According to Ross Dawson, a globally recognized futurist and a strategy advisor, "We are now at the opening phases of what is a global talent economy. Talent is now everywhere and far more available. We're seeing professionals increasingly working independently rather than necessarily in large corporations; we are seeing retired people who are interested in continuing to be engaged and entrusted to projects. And, clearly, we have access to people around the world. So, we are moving from a world where the talent was all inside big organizations to a very fluid world where the talent is available globally. And there is now a whole host of tools and platforms to be able to access all of this talent in a wide variety of ways" (Davey, 2010). It is beyond doubt that crowdsourcing is here to stay and that it could help us in organizing better events. Yet we also agree that whatever we do, we need to work out a good strategy for how to be most efficient and successful. In his book *Crowdsourcing*, Jeff Howe (2009) distinguishes four primary strategies that can also be applied to events. They are based on the four different types of crowdsourcing we have already discussed earlier.

- **Crowd Voting.** According to Howe, crowd voting is the most popular form of crowdsourcing because it generates the highest levels of participation. Crowd voting is also used as a marketing tool to develop contest campaigns.

- **Crowd Wisdom.** According to Howe, "Given the right set of conditions, the crowd will almost always outperform any number of employees, a fact that many companies are increasingly attempting to exploit." The best example of this type of crowdsourcing strategy could be Wikipedia, a tool that is used by thousands of people all over the world every day.

- **Crowdcreation.** This can be defined as activities such as asking individuals to film TV commercials, perform language translation, or solve challenging scientific problems. This strategy is combined with crowd voting to create marketing campaigns.

- **Crowdfunding.** This describes the collective cooperation, attention, and trust by people who network and pool their money and other resources together to finance or help individuals or groups who usually cannot finance their projects through other means.

Thus, it is important to decide what our aspirations are and weave crowdsourcing strategy into the whole planning process of the event.

After you have created your crowdsourcing strategy, you have to choose the tools that you will use to execute your strategy.

One of the greatest benefits in using crowdsourcing is that you stimulate the competition of ideas and solutions; you create an environment in which a collective effort is a must, in which everyone can learn new things.

Using social media for your crowdsourcing tasks in the planning process of your event can be really creative, fun, and effective. You will just need to find the most efficient tool for each crowd. We cannot possibly discuss the whole range of available tools in detail. Nonetheless, we will take a closer look at some of the currently most popular ones.

TWITTER

Needless to say, what you definitely have to bear in mind in case you decide to use Twitter is to create a unique hashtag (#) to explain clearly what you want to accomplish and to have followers, who will be tweeted. In Figure 14.2 you can see the process of using

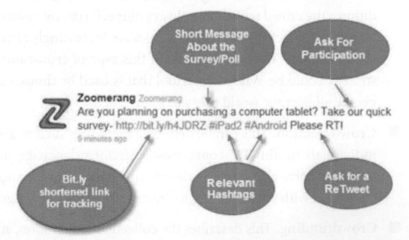

Figure 14.2. Crowdsourcing on Twitter (Source: Zoomerang, part of SurveyMonkey Inc., Palo Alto, California, USA, www.surveymonkey.com, Used with permission.)

crowdsourcing for surveys on Twitter. As we have already discussed earlier, it would be reasonable and useful to offer some kind of incentives to those who will participate. Actually, that is the missing thing in Figure 14.2.

FACEBOOK

If we are looking for a big crowd, Facebook is, certainly, the right place. As of May 2014, Facebook has as many as 900,000,000 estimated unique monthly visitors (eBizMB, 2015). Thus, we cannot escape the network if our goal is to reach the biggest possible crowd in the world. Relying on our previous experience and research, we

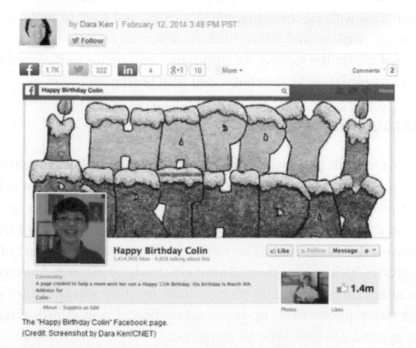

Figure 14.3. Facebook Crowdsourcing (Source: Kerr, 2014)

have arrived at the conclusion that crowdsourcing with consumers can result in greater customer satisfaction and can increase in a positive way the word-of-mouth contacts among participants. It is nearly impossible to hold an event today without creating an event on Facebook. You can use different applications to ask questions, share ideas, search for solutions, etc. For example, *The Wall Street Journal* asked its readers on Facebook to share what they considered interesting and exciting in the job market and highlighted some of the submitted stories in an article. Among the most popular tools used on Facebook, the following ones stand out: contests, polls, and surveys (Lavrusik, 2011).

PINTEREST

The fact that Pinterest counts on pictures organized in sections makes it easy to attract new people to your cause. You have to be aware, however, that a female population dominates Pinterest. Using Pinterest you can find out how your event fits into the lives of your guests and gain important insights into improving your creative strategy or event positioning. As with all crowdsourcing activities, you have to focus more on being inspirational rather than on technical factors, as people in a crowd are on the lookout for inspiration and opportunities to show their knowledge, not to be confronted with boring stuff.

INSTAGRAM

Instagram is the place where smartphone users publish their pictures. So if pictures are part of your crowdsourcing efforts, Instagram is the right place for you. Users can follow other users, which in turn can be followed as well. Pictures can be "liked" and commented on. It is a well-known fact that people love to watch the activities they are part of in progress, which can be achieved most effectively by pictures. Moreover, Instagram users are mainly young adults. In other words, if they are your target, photos make up an important part of your crowdsourcing strategy and you have to be on Instagram.

STEP 1:
Be sure that you post your question to the Public audience

STEP 2:
Make your long URL short and easy measure

STEP 3:
Ask your audience for help, saying clearly what you want from them

Figure 14.4. How to Use Google Plus for Crowdsourcing

GOOGLE PLUS

There are 300 million active users in the Google + Stream. Google+ user base comprises roughly 60% males and 25% females as of November 2013 (Imparture, 2014). As we know quite well, all our crowdsourcing efforts are aimed at reaching the maximum number of people, thus we have to know how to reach that goal. In Figure 14.4 you will find some tips in using G+. A very important step is to add people who responded to your question to a circle in which you will be sharing the results from the crowdsourcing question. That will make them feel personally connected to your post, having contributed to it in some way, and will hopefully help you get the word out.

YOUTUBE

In short, the video is the king of the content movement nowadays, surpassing everything else in its impact on people. With this in mind, you have to consider and plan carefully in advance how to use video content for your crowdsourcing efforts. There are two main ways: Sharing by video your call for action or encouraging people to create video that will meet your needs. And, of course, it could be a combination of both —asking people for ideas and executing those ideas in video format.

Just remember the "Happy" video wave, and your imagination will help you come up with plenty of great ideas.

SLIDESHARE

Presentations are the real standard for business today. Using and sharing presentations could become a convenient and efficient contribution to your crowdsourcing efforts.

SCRIBD

Provided that people need to read various materials before becoming part of your crowdsourcing efforts, using Scribd might be a smart decision. Scribd is a digital library featuring a subscription service with premier books including *New York Times* bestsellers and classics.

It will not be possible here to take a closer look at all the platforms that you can use in practice. However, you have to be curious to find new platforms and what people can do with them.

Although social networks can be helpful, don't forget that there are numerous other websites you can use to work efficiently in collaboration, among which Evernote, Google Drive, Get Satisfaction, Delicious, and Dropbox are worth mentioning.

The process of crowdsourcing is now easy and convenient because of the huge penetration of Internet. Figure 14.5 presents the scheme of the process. In this process there is a wide range of platforms that can help you become more efficient.

Figure 14.5. The Process of Crowdsourcing

We cannot possibly encompass all online platforms which you can use in your practice. Besides, we are sure that when you read this section of the book, there is bound to be something new online for you. Nevertheless, we will touch upon some of the currently most popular crowdsourcing tools to provide an overall impression of some possible expectations:

- **crowdSPRING** (crowdspring.com) helps people from around the world get access to creative talent, as well as all those with a more a creative streak to find new customers across the globe. In other words it is an online marketplace for crowdsourcing logo designs, web designs, graphic designs, industrial designs, and writing services.

- **99designs** (99designs.com) is an online graphic design marketplace that enables customers to source graphic design work quickly.

- **oDesk** (odesk.com) is a global job marketplace with a series of tools targeted at businesses that intend to hire and manage workers from distant and remote backgrounds.

- **Microworkers** (microworkers.com) is a platform by which you can crowdsource your micro jobs to more than 500,000 workers worldwide.

- **Innocentive** (innocentive.com) is a global platform in crowdsourcing innovation problems set up for the world's most educated and intelligent people who compete to provide ideas and solutions for important business, social, political, scientific, or technical challenges. Its global network of millions of problem solvers, proven challenge methodology, and cloud-based technology are combined in order to help clients transform their economics of innovation through rapid solution delivery and the development of sustainable, open innovation programs.

As you may assume, crowdsourcing is a big phenomenon. Crowdfunding is part of that new movement. Although the practice of crowdfunding is not completely new for the event industry, the concept is still formally recognized as a new industry by numerous consumers. Crowdfunding is by definition "the practice of funding a project or venture by raising many small amounts of money from a large number of people, typically via the Internet." If we have to adapt the definition to the event industry, it will look as follows: It is, in fact, the practice of funding an event by raising many small amounts of money from a large number of people, typically via the Internet. What is more, buying tickets or paying your fee in advance as well

Table 14.2. Top 10 Crowdfunding Sites by Traffic Rank (Source: http://www. crowdfunding.com/ 24 of April 2016)

Rank	Crowdfunding Site	2014 Volume	US Alexa Rank	Fee	Important to Know...
1	gofundme.com	$470M	321	5%	Over $2 Billion raised for personal fundraisers. Processing fee of 2.9% + $0.30 applies.
2	kickstarter.com	$444M	225	5%	Personal fundraising not allowed. Creative only. Processing fees of between 3-5% apply.
3	indiegogo.com	???	554	5%	3% processing fee. $25 fee for international wire.
4	teespring.com	1,170	823	10%+	T-shirt crowdfunding site. Fees vary based on t-shirts selected for sale.
5	patreon.com	1,057	2,378	5%	Must pledge an on-going amount. Creative projects only. Additional processing fee of 4%
6	youcaring.com	2,767	2,110	5%	5% fee is suggested to campaign donors. Processing fee of 2.9% + $0.30 applies.
7	crowdrise.com	2,898	4,703	5%	Free accounts charge 5%, paid accounts are 3%. Processing fee of 2.9% + $0.30 applies.
8	Donorschoose.org	4,685	6,467	15%	Optional 15% fee to support DonorsChoose.org. Donation is 100% tax-deductible.
9	kiva.org	5,418	5,946	15%	15% fee is suggested to campaign micro-lenders. Processing fee of 2.9% + $0.30 applies.
10	Giveforward.com	16,448	4,169	5%	5% fee is charged to campaign creators. Processing fee of 2.9% + $0.50 per transaction applies.

as donating money for events are practices that we have been familiar with for long time. One might wonder what is new about it. Of course, the great novelty is the Internet along with the great variety of opportunities available to us that will help us improve the quality of events. If we go back to the 5W questions, the most important query for crowdfunding is **Why.** People will be motivated to fund our event if they care not so much about what you do, but why you do it. By focusing on the bigger, long-term goals, the driving force behind an event, we will be able to create a unique community of like-minded individuals.

According to Tanya Prive (2012), contributor at *Forbes*, there are three main reasons why people unconnected to a project or business would support it:[5]

1. They connect to the greater purpose of the campaign

2. They connect to a physical aspect of the campaign, like the rewards

3. They connect to the creative display of the campaign's presentation

If we replace the word campaign with event, we will get a brand new sound basis for our crowdfunding strategy in setting up an event:

1. They connect to the greater purpose of the event

2. They connect to a physical aspect of the event, like the rewards

3. They connect to the creative display of the event's presentation

Following these guiding rules and using some of the top crowdfunding platforms, you can organize and execute an impressive, purposeful event.

In fact, all these sites create an online community of people who care about your cause (event) and are ready to follow your progress and help you in achieving your goal. And, in fact, that means much more than receiving just financial support.

Jason Graf (2013) from CrowdIt in an article published in *Business News Daily* outlines 15 simple ways to increase the odds of crowdsourcing success. Among them are the use of video, offering pledges which secure 20–30% of the desired funding, setting realistic goals, creating a great pitch, creating a page, showing who you are, sharing as many details as possible, showcasing unique goods, and offering rewards and incentives. Not forgetting the fact that it should be personal will help you develop a personal approach and get used to communicating frequently with the crowd. It's important to be passionate about your event and not to give up. As with all events, it is important to keep your promises and commitments, express your gratitude to those who have helped you, and share what you have achieved together.

According to statistics, a typical donation is $75 on average (Graf, 2013).

It is important to point out the significance of crowdsourcing and crowdfunding principles, which have a lot in common with your events. The following may serve as an example of how the above-mentioned factors can transform events and make them gain an ever wider scope.

We would like to draw your attention to the phenomenon called "crowd-accelerated innovation". This term was coined by Chris Anderson (2010), the curator of the TED Conference in 2002. In a TEDTalk he talks about how he "became intrigued by a different way of thinking of large human crowds, because there are circumstances where they can do something really cool. It's a phenomenon that any organization or individual can tap into." To use this phenomenon effectively, you need three things. Think of them as three dials on a giant wheel, and when you turn up the dials, the wheel starts turning. Those things are:

- **Crowd**. A group of people who share a common interest. The bigger the crowd, the more potential innovators there are.

■ **Light**. This refers to a clear, open visibility of what the best people in that crowd are capable of, because that is the way to learn how to gain the necessary power to participate effectively.

■ **Desire**. With the absence of desire, nothing is bound to happen.

We have already examined closely the first two elements. As regards the third one, desire, we have touched upon it only in the extract concerning rewards. Yet one must not forget that desire is of prime importance, as it proves why people will want to become part of the crowd. The secret of desire lies in your answer to the questions Why and Who. Remember that people want to be part of something big, something they believe in, and, last but not least, to be in a great company. To sum it up, without desire the crowd will not respond.

If desire is the crucial ingredient about a crowd, light is the tricky part from the point of view of event organizers, "because it means you have to open up, you have to show your stuff to the world. It's by giving away what you think is your deepest secret that maybe millions of people are empowered to help improve it" (Anderson, 2010).

TED has two successful initiatives in crowdsourcing. First, by opening up their translation program, thousands of volunteers have translated TEDtalks into more than 70 languages. The latter makes videos from their conferences easy to watch all over the world. The word-of-mouth phenomenon is out there, as people translate and talk to their friends about what they have done and why. And second, by giving away the TEDx brand, they suddenly have a thousand-plus live experiments in the art of spreading ideas. In May 2014 TED celebrated its 10,000th event, TEDx! Since its start in 2009, TEDx events have been held in as many as 167 countries, at an average rate of eight per day. Imagine this —8 events per day done by a great crowd of people willing to help you spread ideas.

You have certainly watched at least one TEDTalk. And if a TED conference can tap on the power of the crowd in such a splendid way, why not at least give it a go?

CONCLUSION

At the end of his first article about crowdsourcing Jeff Howe wrote that the term "crowdsourcing owes a debt to James Surowiecki's eye-opening and entertaining book, *The Wisdom of Crowds*. But as Surowiecki, himself notes in that book, we both owe a further debt to Charles Mackay's *Extraordinary Popular Delusions and the Madness of Crowds*, a work from the early Victorian era that should be a required reading for anyone interested in the subject" (Surowiecki, 2005; Howe, 2006). If you want to get a deeper insight into the idea of crowdsourcing and crowdfunding, a great starting point is to read those two books. If you ask yourself what the crowd can do for you, you are heading in the wrong direction. You have to ask yourself what you can do for the crowd.

DISCUSSION QUESTIONS

1. What will motivate you to become part of a crowd?

2. What other examples of a crowdsourcing event, except TED, can you list?

3. What are your attitudes to crowdsourcing as a power in modern society?

4. How would you crowdsource a local new sports event?

TASK

Try to become part of a crowdsourcing effort in which you are interested. Visit the platform we discussed here and take part in a task. What do you think? What have you learned that can be useful in making events and using the power of the crowd to do better events?

REFERENCES

Anderson, Chris (2010). How web video powers global innovation, Ted.com https://www.ted.com/talks/chris_anderson_how_web_video_powers_global_innovation?language=en, accessed 10.05.2014 July 2010

Brabham, Daren C. (2008). "Crowdsourcing as a Model for Problem Solving", Convergence: The International Journal of research into new Media Technologies, 2008, Sage Publications

Creamer, Matthew (2008). "Obama Wins! ... Ad Age's Marketer of the Year," Advertising Aging, October 17, 2008. http://adage.

Davey, Neil, (2010). "Ross Dawson: Six tools to kickstart your crowdsourcing strategy," http://www.mycustomer.com/marketing/strategy/ross-dawson-six-tools-to-kickstart-your-crowdsourcing-strategy, accessed 10.05.2014 July 1, 2010

eBizMBA Inc. (2015). Top 15 Most Popular Social Networking Sites. October 2015. Accessed October 15, 2015. http://www.ebizmba.com/articles/social-networking-websites

Howe, Jeff (2009). Crowdsourcing: Why the Power of the Crowd Is Driving the Future of Business, Crown Business, 2009

Howe, Jeff (2006). Crowdsourcing: A Definition, June 02, 2006, http://crowdsourcing.typepad.com/cs/2006/06/crowdsourcing_a.html, accessed 10.05.2014

Goldblatt, Joe (2014). Special Events: Creating and Sustaining a New World for Celebration. Wiley

Graf, Jason. "15 Simple Ways to Increase Odds of Crowdfunding Success." Business Daily News. August 28, 2013. http://www.businessnewsdaily.com/5009-crowdfunding-tips-tricks.html, accessed 10.05.2014

Hovnanian, Stephan (2013). Google Plus Crowdsourcing the Right Way (Source: http://www.websighthangouts.

com/crowdsourcing-on-google-plus/) Posted: April 30, 2013

Imparture. 2014. Google+ Marketing Playbook. https://www.imparture.com/uploads/files/2014/12/googleplussample.pdf

Kerr, Dara (2014). "Facebook crowdsourcing gives lonely boy 1.4 million friends," CNET. http://www.cnet.com/news/facebook-crowdsourcing-gives-lonely-boy-1-4-million-friends/ February 12, 2014

Lavrusik, Vadim (2011). Using Facebook to crowdsource a story. Journalists on Facebook. August 31, 2011 at 3:58pm. https://m.facebook.com/notes/journalists-on-facebook/using-facebook-to-crowdsource-a-story/264392416906113

Lutz, Monte (2009). Barack Obama Social Media Toolkit, http://cyber.law.harvard.edu/sites/cyber.law.harvard.edu/files/Social%20Pulpit%20-%20Barack%20Obamas%20Social%20Media%20Toolkit%201.09.pdf, accessed 10.05.2014

Miller, Jason (2011). "How to Use Twitter for Crowdsourcing and Simple Market Research," SocialMediaToday.com http://www.socialmediatoday.com/content/how-use-twitter-crowdsourcing-and-simple-market-research.

Prive, Tanya (2012). What Is Crowdfunding And How Does It Benefit The Economy, http://www.forbes.com/sites/tanyaprive/2012/11/27/what-is-crowdfunding-   and-how-does-it-benefit-the-economy/, accessed 10.05.2014

Surowiecki, James (2005). The Wisdom of Crowds: Why the Many Are Smarter Than the Few and How Collective Wisdom Shapes Business, Economies, Societies and Nations, Anchor, 2005

ADDITIONAL RESOURCES

Abrahamson, Shaun, Peter Ryder, and Bastian Unterberg (2013). Crowdstorm: The Future of Innovation, Ideas, and Problem Solving, Wiley.

Brabham, Daren C. (2013). Crowdsourcing, MIT Press Essential Knowledge.

Gansky, Lisa (2012). The Mesh: Why the Future of Business Is Sharing, Portfolio Trade

Hinchliffe, Dion (2009). Crowdsourcing: 5 Reasons It's Not Just For Startups Any More. Ebiz. September 25, 2009. http://www.ebizq.net/blogs/enterprise/2009/09/crowdsourcing_5_reasons_its_no.php

Mackay, Charles (1999). Extraordinary Popular Delusions & The Madness of Crowds. Wordsworth Edition

Part IV

The Future of Meeting and Events Technology

CHAPTER 15

Futuristic Tech Trends That Will Influence Our Events

> *"It is an interesting feature of cultural change that, for a period of time, new technologies tend to be used to do the same old thing."*
>
> —*Sir Ken Robinson,*
> *an English author, speaker, and international advisor on education to*
> *government, non-profits, education, and arts bodies (2014)*

LEARNING OUTCOMES

As a result of reading this chapter, you will learn how to:

- Be aware of the new trends impacting meeting and events technology

- Research information about new trends in meeting and event technology

- Forecast new applications of technology

- Search for new and emerging technologies

- Continuously improve your meeting and events outcomes through adoption of new technologies

INTRODUCTION

When meeting and event management was first developed within higher education, we understood that a special event has clear beginning and clear end. However, since the development of the Internet, social media, and networks has become an integral part of everyday life, we can say that this knowledge is no longer true. As a result of global technological networks, events today are without end. Events are indeed now a never-ending story.

At one time we were taught that the successful event has to create a community of people and to have positive social impacts upon the environment. At that time it was a huge challenge for us, as for a community to endure, we needed channels for communication. During that time, communication was slow and expensive. Now to communicate with huge groups of people, you need just a device and an Internet connection. We can say that one of the biggest challenges in event management can be more easily overcome. But the second one—to create events with a more positive impact upon society—remains a difficult challenge for every event manager.

Regardless of the mass penetration of technologies in our life, we still need face-to-face communication with people. This is confirmed by a study of 7,000 meeting and event planners who stated that their preference was to blend online with face-to-face learning to create better outcomes (McEuen and Duffy, 2010).

The data collected from organizations such as Meeting Professionals International and the Professional Convention Management Association (World Tourism Organization, 2014) clearly shows that the number of organized events is not just stable but is growing. Events are a permanent fixture in our lives. They have been an integral part of all human history and will be with us forever. That's not to say that events didn't change in the last decade.

Events have changed significantly, and many more changes will enter the field in future decades. What exactly will change? No one knows for sure, but by analyzing what happens we can forecast and then to see if they will became a reality. Here we will focus our attention only on those possible changes that are connected with technology.

For example, in the year 2000, Professor Joe Goldblatt (2013) delivered a lecture in Sydney, Australia, describing his forecast for events in the new millennia. Goldblatt forecast that the majority of meeting and event registrations would be transacted online in the first decade of the twenty-first century. This forecast was accurate. He also forecast that by 2015 event planners would be broadcasting events from planets besides our own planet Earth. While this has not been achieved, it may indeed occur in the future as national governments continue the space race to Mars and other planet. Finally, Goldblatt forecast that in the year 2025 many events would have full robotic capacity, thereby reducing the need for some human labor. Once again, while robotics has significantly developed in recent years, we will not be able to confirm this trend until these devices are more generally utilized and adapted by the meetings and events planners of the future.

If we try to focus on all the new meeting and event technologically friendly gadgets, this chapter will be out of date before you finish it.

However, there are some emerging and new technological tools that merit further examination.

We will start with Google Glass. There are already plenty of applications popping up all over the place for this incredible pair of glasses. There are yet numerous applications of Google Glass at events. The two main applications for our industry are in ticketing and livestreaming.

Event staff can immediately scan event tickets by wearing Glass or attendees can scan an intelligent code to get clearance. Secondly live-streaming is the most logic use of Glass, thanks to hangout integration.

You can list many more applications.

In the *60 Minutes* television show on the American television network, CBS, Amazon CEO Jeff Bezos revealed that Prime Air delivery by drone could be a very real possibility in the not too distant future (Rose, 2013). A drone is an aircraft that has the capability of autonomous flight. A drone can follow a mission from point to point without a pilot on board. Drones are typically guided by Geographical Positioning Systems (GPS), but soon this will also be possible through vision and other sensors. With a drone we can explore the limits of our imagination to create a better experience for our guests through helping manage deliveries, observations, and even monitoring crowd control.

Crowd management and control is one of the most important issues in the meetings and events industry. We do not have precise technology that empirically provides details regarding crowd movement. The Internet of Things (IOT) is a technology trend that has had little impact so far on events. Google is at the forefront of IOT innovation. With IOT we can measure temperature, humidity, light, pressure (including nearby footfalls), motion, air quality, and both RF (radio frequency) and ambient noise. All of the data is sent back at intervals of 20 seconds or so, collected by Google's App Engine, with analysis performed by its BigQuery Big Data analysis tool.

It is not possible to list all of the new and emerging technologies that will impact the meetings and events in the future. The list will not be up to date when you read it. Instead we will try to focus on some of the major trends connected with these technologies.

THE HYBRID STANDARD

Did you know that Ferdinand Porsche developed the first gasoline-electric hybrid automobile in the world in 1901? It took nearly a century for the hybrid-electric vehicles to become widely available. In 1997, we saw in Japan the introduction of the Toyota Prius, followed by the Honda Insight in 1999.

One major trend in meetings and events is the hybridization between the face-to-face (live) and virtual (online) experience. Many meetings and events industry associations have demonstrated through online as well as face-to-face registrations that hybrid meetings and events are indeed a major trend.

- **Hybrid meetings will be an important part of the industry's future.** In the MPI study, seventy percent of respondents felt that hybrid meetings would be important to the future of meetings, though further analysis reveals that meeting professionals are still becoming familiar with the medium.

- **The hybrid meetings movement hasn't gained critical mass.** Fifty percent of the MPI respondents have never organized a hybrid event, and another 25 percent have never attended or even helped to organize a hybrid event.

- **Technology is not the only factor in the success of a hybrid event.** While many meeting professionals cite technology as a barrier to the success of a hybrid event, others also point to people, processes, and formats.

We may argue that hybrid events are a permanent fixture of meetings and events. Through hybridization, attendees receive more information before, during, and following the meeting or event, and online events are not as rare as they were five years ago. As part of our future ability to harness and expand the power of meeting and event management, we will have to further develop our knowledge of hybrid events and learn how to plan, organize, and evaluate them.

LOCAL MEANS GLOBAL

Thanks to new technologies every single local event now has the potential of going global. That is a huge change as we are used to think about our local events as only local and dedicated to the local community. Now they are going global and can reach the global community. With this in mind, we have to think simultaneously about local aspects and global reach. This shift has to happen first in our mindset, and after that we can find ways to became part of every single local event. What does it means in meetings and events management?

Here are a few key technological considerations:

- Language and translation of information about the event
- Compliance with different cultures and traditions
- Being open to different points of view
- Planning in accordance with time differences
- Being aware that local publicity can quickly became global news.

Of course, millions of local face-to-face meetings and events will be conducted in the future. Local meetings and events are part of our culture and heritage, but now they can and will reach global audiences via the Internet. Guests will share opinions, photos, and experiences with their global networks in different social platforms. If you manage the globalization of local events, you can achieve better results and reach people all over the world.

MOBILE LIVING

How often do you use mobile communications devices such as smartphones and tablets for Internet access as compared to desktop computers? More and more we prefer to use the mobile devices in our pocket. According to Morgan Stanley, mobile web usage has significantly increased over traditional desktop web browsing between 2007 and 2015.

So what does that mean for meetings and events management? We have to adapt our stereotype of online communication to a new mobile reality. We have to make the transition in our mentality from desktop communication to mobile communication, and we need to use different applications for creating a better experience for our guests.

AUTHENTIC EXPERIENCES

3D movies and television shows are increasingly more prevalent. The consequences of that 3D experience are that we want to have authentic experiences in all our activities. We want to see, hear, feel, smell, taste, and touch to gain a new experience of the world around us. Today augmented reality is the new thing in tech and media circles. It's expensive, and few events organizers can afford it, but soon it will be an integral part of a different kind of events. We can expect that the number in front of the D will grow, and we will want to have 4D, 5D, etc. technologies. There are many applications that help us to use augmented reality to create a better experience for an event's guests. The challenge for us is to understand the technology, to understand the needs of our guests, and to find the perfect match between our intentions to use newest technologies and to keep the focus on the personal experience of each single guest.

SUSTAINABLE MEETINGS AND EVENTS WITH TECHNOLOGY

In 2012, after the hard work of experts from all over the world, the first ISO certificate for environmental sustainability in event management was publicly announced. ISO 20121:2012 event sustainability management systems are only one of the signs that people are more and more concerned with the environment and they want to consume ethically. New technologies can help event organizers to reduce waste, to be friendly to environment, and to be concerned with what footprint is left after the event ends.

The global trend of sustainability is changing the practice of events all over the world, and new technologies are here to help us make sustainable events. Simply greening our event is not enough. People expect more from event organizers. Good practices have become known globally, and we have to be aware of the best cases and try to move to the highest standard in our profession.

EVENTS AS SOCIAL CAPITAL

What does "social capital" mean? The central premise of social capital is that social networks have value. Social capital refers to the collective value of all "social networks" (who people know) and the inclinations that arise from these networks to do things for each other ("norms of reciprocity"). Thanks to social media and networks, we can share what we are doing, where we are doing it, with whom, etc. Smart event managers have to focus on the opportunity to convert their events in the important currency of social capital. According to Future Foundation (2014), "The forthcoming decade will be defined strongly by the ability to accumulate social capital." People share information using their personal networks. They choose what to share and with whom. Events managers have to be aware of the importance of social capital and to use their imagination in combination with new technologies to transform and create events to take advantage of the important currency of social capital.

CO-CREATED EVENTS

When using new technologies, consumers not only consume products and services; they also create the reputation of the brand. The same is true for the reputation of events. We have discussed the crowdsourcing practice. In the future new technologies will help people to become a more integral part of events, and we have to be prepared for that kind of cooperation.

DISCUSSION QUESTIONS

1. What are some of the obstacles to conducting hybrid meetings and events?

2. What do you believe are some further demographic and social trends that will impact the future development of technological solutions for meetings and events?

3. What new technologies will be needed to improve the evaluation of meetings and events in the future?

TASK

Work with a small team of students to develop a plan for a new mobile application for the meetings and events industry that will increase attendance and improve the overall experience for your attendees.

REFERENCES

Future Foundation (2014), accessed http://futurefoundation.net.

Goldblatt, J. (2000) Events Beyond 2000. Sydney, Australia: University of Technology Sydney.

Goldblatt, Joe (2013). "A Future for Event Management: The Analysis of Major Trends Impacting the Emerging Profession." http://sm.avito.nl/wp-content/uploads/2013/02/Goldblatt-J.-2000.-A-future-for-event-management-the-analysis-of-major-trends-impacting-the-emerging-profession.pdf

McEuen, Mary Beth, and Duffy, Christine. The Future of Meetings: The Case for Face-to-Face. The Maritz Institute White Paper, September, 2010. http://www.themaritzinstitute.com/perspectives/~/media/files/maritz-institute/white-papers/the-case-for-face-to-face-meetings-the-maritz-institute.pdf

Robinson, K. (2014). http://sirkenrobinson.com, Accessed 11.25.2015.

Rose, Charlie. Interview with Jeff Bezos, 60 Minutes, December 1, 2013. http://www.cbsnews.com/news/amazon-unveils-futuristic-plan-delivery-by-drone/

World Tourism Organization (2014), AM Reports, Volume seven – Global Report on the Meetings Industry, UNWTO, Madrid. http://www.imexexhibitions.com/media/350548/UNWTO_meetingsindustry_am_report%20(2).pdf

ADDITIONAL RESOURCES

Beyond 2000 (2000) Sydney, Australia:Australian Centre for Events Management, accessed 5, July 2014. Viewed: http://epress.lib.uts.edu.au/research/bitstream/handle/2100/430/Proceedings%202000%20Conference.pdf?sequence=1

Convention Industry Council, Economic Significance Study, accessed: http://www.conventionindustry.org/ResearchInfo/EconomicSignificanceStudy.aspx

Future Trends Impacting the Exhibitions and Events Industry, IAEE White Paper by Francis J. Friedman, http://www.iaee.com/downloads/14 04201765.17224000_1eb8041bd7/IAEE%20Future%20Trends%20Impacting%20the%20Exhibitions%20and%20Events%20Industry%20White%20Paper.pdf, accessed at 15.06.2014

A Futurist's Thoughts on Consumer Trends Shaping Future Festivals and Events, Ian Yeoman, International Journal of Event and Festival Management, Vol. 4, Iss:3, pp. 249-260.

Hybrid Meetings and Events, Jenise Fryatt, Rosa Garriga Mora, Ruud Janssen, CMM, Richard John, Samuel J. Smith, http://www.mpiweb.org/docs/default-source/Research-and-Reports/HYBRID-Executive_Summary.pdf, accessed at 15.06.2014

Meetings and Conventions 2030: A study of megatrends shaping our industry, http://www.icahdq.org/images/newsletter/newsletterpdf.pdf, accessed at 15.06.2014

Meeting and Event Technology Case Studies

CASE #1: CRASHING WI-FI AT FIRST PAPERLESS MEETING

Ten thousand biologists attended a scientific conference in Barcelona, Spain. This conference was one of the first to provide a paperless experience for the delegates. All of the schedule, social event information, abstracts and other critical information would be provided from the cloud. However, within the first ten minutes of the meeting the conference centre information services manager told the meeting planner that their Wi-Fi system could not handle 10,000 persons logging on simultaneously, and their system had indeed crashed.

The Challenge

When the delegates could not access their critical information from the cloud, chaos ensued. Hundreds of delegates stood in long lines at the registration and help desks asking for paper copies of the schedule, abstracts, and social event information. The meeting planner received thousands of irate e-mails. Various scientific online chatrooms we filled with posted complaints about the meeting.

The Opportunity

The meeting planner could have had a backup system in case of systemic failure. One potential backup was to collect e-mails and mobile numbers from all delegates and then send them messages directing them to alternative sites to collect their information.

How This Case May Be Applied to Advance Your Future Career

You should meet with the venue information services manager and discuss back up plans in case the Wi-Fi system fails. You should have this issue highlighted in your risk management plan. When interviewing for future jobs in the meetings and events industry, you should describe how you would handle disruptions in technology before, during, and after your meeting or event.

CASE #2: FAILURE OF ELECTRONIC TICKETING SYSTEM

A new electronic ticketing system has been installed for a major European arts and cultural festival. Within a few hours of usage, the system fails and there is massive confusion as tickets are not printed, financial transactions are not conducted, and customers are extremely irate as they are missing the performances they wish to attend.

The Challenge

The new ticketing system did not have a redundant platform to replace the faulty original system. As a result, ticket buyers had to be issued handwritten paper tickets or in some cases had their hands stamped with a rubber stamp after paying for their tickets with cash. This led to massive errors regarding cash handling and a 25 percent decline in box office transactions.

The Opportunity

The arts and cultural festival organizers should have used a redundant system approach so that if there was a failure they could switch to plan B to satisfy the needs of their guests. The original system should have undergone and extensive period of live testing prior to implementation.

How This Case May Be Applied to Advance Your Future Career

When applying for a new job, describe how you would select, develop, test, and implement a new online registration or ticketing system. Describe your plan for creating a redundant system and how you will communicate in real time with attendees and ticket buyers so that they are aware of how to complete their transaction. Describe how you will practice due diligence when appointing a registration or ticketing firm to provide services for your meeting or event.

CASE #3: DELAYED ELECTRONIC TICKET DELIVERY

Millions of tickets are placed on sale for a mega sports event. Within a few days, ticket purchasers are receiving messages saying that due to an overwhelming response for tickets, there will be delays. Ticket purchasers fear that they will not receive the tickets they desire and begin to e-mail the organizer and also post negative messages on social media sites.

The Challenge

One of the major challenges with selling large numbers of tickets is the wide range of broadband capacity of the potential buyers. The ticket seller has little to no control over individual broadband width and capacity and this may slow the dissemination of tickets.

The Opportunity

Ticket buyers should be informed in advance of the challenges of broadband capacity and offered alternative ways to obtain tickets or to register. Alternatives may include telephone, e-mail, in person at box office, and even through text messaging.

How This Case May Be Applied to Advance Your Future Career

When applying for a future meeting or event job, let your employer know that you have a wide range of effective solutions to distribute tickets or collect registrations for your event. Describe each solution and compare and contrast the benefits for the meeting and event planning organization as well as the participants.

CASE #4: SECURITY FAILURE OF PERSONAL ELECTRONIC DATA

The hard drive of your association headquarters has been hacked, resulting in the massive theft of thousands of individual records, including health information about your members.

The Challenge

One of your staff members is married to a member of the media and mentions this theft to her husband who quickly informs his editor, and a story soon appears on page one of the local newspaper.

The Opportunity

It is impossible to provide 100 percent security when using online systems. Therefore, you must develop a risk aversion, risk management and risk control policy to protect the data entrusted to your organization. This policy must identify key leaders who will serve as stewards of this information and when there is a breach of security implement the procedures identified through your policies.

How This Case May Be Applied to Advance Your Future Career

When applying for a job, tell your interviewer how important it is to protect the information entrusted to your organization. Describe examples of policies that you would follow or implement to reduce risk associated with data protection.

CASE #5: NEGATIVE ELECTRONIC FEEDBACK ON YOUR MEETING VENUE

TripAdvisor.com suddenly has a large number of negative reviews posted about your upcoming convention hotel. As a result, your organization is receiving a number of e-mail queries from potential registrants, and some current attendees are asking for refunds from the hotel and desire to book rooms outside of your contracted block.

The Challenge

The individuals posting negative reviews on TripAdvisor.com are leisure tourists whose individual complaints have now gone viral as other guests receive e-mail notices about this hotel property. You must quickly ensure that accurate information is being disseminated to your current and potential attendees.

How This Case May Be Applied to Advance Your Future Career

When applying for a future job, discuss the importance of third-party social media referral sites such as TripAdvisor.com and describe how you would organize a group of your meetings and events participants to refute negative reviews on these sites. Also describe how you will effectively communicate with your future and current participants to ensure them that you are investigating these complaints and will report back with your findings in a reasonable amount of time. To reduce the number of persons who attempt to book outside your convention hotel block, remind them of the value of staying in the headquarters hotel (increased networking, efficient use of time, your ability to leverage your position and act on their behalf) and the potential penalties for moving outside the hotel (a cash penalty imposed by your association, loss of time when walking to and from the headquarters hotel, loss of key networking opportunities.

CASE #6: LIVE EVENT IDEAS WORTH TRANSLATING

When the organizers of the conference TED created TED.com and uploaded videos from the TEDTalks, the interest was tremendous. The videos were all in English language. Although English is spoken by many people around the world, the sole use of English became a barrier for spreading the ideas globally. The TED community was growing and the idea of translating TEDTalks began to gain interest. Passionate viewers around the world started asking if they could translate talks in order to share them with friends and family. If that could happen, then the Talks from the conference would be able to reach even more people who do not speak English, and they would be inspired by the ideas that TED speakers shared.

The Challenge

The real need and an opportunity to radically open accessibility to TEDTalks by developing a translation system that would allow volunteers to translate their favorite talks into any language. Another challenge was to guarantee the best quality of the translations.

The Opportunity

Making TEDTalks available in many languages, the TED conference promises "ideas worth spreading" and is going one step further. TED launched in 2009 The Open Translation Project. It started with 300 translations in 40 languages, created by 200 volunteer translators. And the heart of the project is that before going public, the translation should be reviewed by a volunteer. TED encourages translators to know each other and to work together for offering better translations to millions of people. Today, more than 50,000 translations have been published in 104 languages (and counting), created by more than 15,000 volunteers. Recently, the project expanded to include the transcription and translation of TEDxTalks, the translation of TED-Ed lessons, the localization of TED's Android app and the translation of content distributed by worldwide partners that help grow TED's global footprint.

How This Case May Be Applied to Advance Your Future Career

Being a volunteer creates enormous opportunities for gaining experience, creating good business networks, and acquiring knowledge of what happens behind the scene at events. Through providing volunteer translation services, you can gain invaluable experience and generate important contacts. The only thing you need is idea worth spreading.

CASE #7: WHEN EVENT TECHNOLOGY EXPECTATIONS ARE HIGHER

A big multinational electronics company announced the first 3D mapping event. The event was advertised broadly and attracted many people who were eager to see with their own eyes the newest technology. Their expectations were huge due to the rapid growth of the Internet.

The Challenge

The temperatures were high and the beginning of the event was not properly design. People were made to wait for more than two hours before the actual 3D mapping started. When the time came, the file drop and the organizers tried to broadcast the movie more than five times with no success. The people viewing the event on the outdoor public square viewed five times the first minute of the movie. The disappointment of the audience started appearing almost immediately on social media.

The Opportunity

When you create huge expectations, you have to be ready to deliver. The world is now a small village, and people know what to expect when attending a professional event. Event organizers have to plan the starting hour more carefully and work closely with the technical staff to overcome the high temperatures and ensure the weather will not be a problem for the computers to broadcast the 3D mapping film.

How This Case May Be Applied to Advance Your Future Career

When organizing an event to present a new technology, be careful in creating expectations. Sometimes is better to be honest and to present what will really work rather than take chances. Sometimes is better to be good, than to be first.

CASE #8: LIVE EVENT COVERAGE ON SOCIAL MEDIA

Every event organizer knows that social media is an integral part of each event. In this case, the maintenance of social media content was left to the youngest in the team who had good knowledge of new social media but did not have experience in creating content during the event.

The Challenge

Today people expect to receive content through social media nearly simultaneously with the happening of the event. That means that depending on the size of the event, you need dedicated staff from your event team who will perform only this function during the event.

The Opportunity

The best in the field don't rely on chance. If you want to create coverage from your event at the moment things happen, you have to prepare everything before the event. That means to write posts for your blog, and add only pictures from the event, or write a post for your Facebook page and add only details from the real event before you publish. You must even plan who will make pictures for social media and at what intervals they will be uploaded.

How This Case May Be Applied to Advance Your Future Career

When thinking about live coverage of an event in social media, all content should be written during the event. With a good preparation before the event, you can have great live coverage. Having time to plan in advance always is a good choice and part of your success.

CASE #9: WHEN TECHNOLOGY ALLOWS YOU TO BE PERFECT

Famous speakers attract attention and often allow you to attract more people to your event. But in many occasions, the speakers are very busy, unable to travel, and their fees are high. Thanks to new technologies, you may invite famous speakers to virtually participate in your event by using teleconferencing technology. This reduces your expenses and gives you the opportunity to have the speaker your audience wants.

The Challenge

Broadband connection is often taken for granted, and there is always the opportunity to have problems with the quality of the signal. When organizing an event with a speaker who will participate via teleconference, it is best to be prepared for the worse possible broadband connectivity.

The Opportunity

Good planning and good preparation are the secret ingredients to a successful event. It will take time, but you should record the speech of the speaker and have the file ready to be used in case of connectivity problems. This gives essential time to your technical staff to work on the problem and to be ready for the Q&A session. If done professionally, it is possible that people in the room will not understand that you had problems at all.

How This Case May Be Applied to Advance Your Future Career

Having a backup plan is always the best preparation. As you know from Murphy's Law, "what can go wrong, most likely will go wrong." The only way to overcome Murphy's Law is to be prepared and have contingency plans in place, such as a pre-recorded version of the speech.

CASE #10: BAD COMMENTS ON A BIG SCREEN

Using Twitter to place comments on a giant conference screen is now a usual practice at many conferences. Most of the guests are eager to share their opinions with friends in different networks. Most event organizers use this opportunity to show on screen the comments from Twitter that are gathered in one place preceded by the hashtag (#) of the event.

The Challenge

New social media allow every single person to express their individual personal views to a wide audience. Unless the comments are moderated, no one is protected from bad comments.

The Opportunity

Before deciding to share all comments on a big screen, think carefully and use a person who will react on the comments immediately and will be in the position to edit the comments appearing on the big screen. That doesn't mean to delete or remove bad comments, but to be selective about which comments to show on the screen, especially to the speaker at the moment he or she is presenting on stage.

How This Case May Be Applied to Advance Your Future Career

Moderation is a very important part of keeping online presence fair and equitable for your speakers and the audience. Even if you do not use one or another social network officially for your conference, that doesn't mean that people there are not talking about you. When negative comments are received about your event, you have to react by answering, by being polite, and by encouraging your supporters to take part in the overall conversation. However, you are not obligated to show all comments on the big screen in the room while the speaker is present.

CASE #11: LIVE NATION SPECIAL EVENTS

Live Nation Special Events is a series of performances held at unique venues across North America. Crowds of thousands attend each live event that offers extraordinary programming.

The Challenge

Prior to implementing online cloud–based event diagram software, Live Nation's venues found their diagramming abilities challenging as the application they use is so difficult to use that some partners/suppliers chose not to utilize it at all.

The Opportunity

In an effort to overcome that challenge, event venues need to find a new and easy solution. Many venues are now partnered with online-based event diagram design companies. Such adoption can allow venue managers to create multiple tailored diagrams with minimal time and efforts. Event sales professionals for venues can create professional diagrams and provide effective presentations to prospective buyers while saving an immensurable amount of time and spend more time and focus on sales efforts.

How This Case May Be Applied to Advance Your Future Career

Event technology is adopted at increasingly fast rate at most event management functions, including floor design. Today most event venues provide online-based floor diagrams that include detailed specifications in 2D (e.g. dimension, maximum capacity per room setup) and 3D (virtual tour of a venue).

CASE #12: EVENT SALES DURING A CONVENTION HOTEL RENOVATION PROJECT

During construction of a convention hotel, sales staff want to present their clients with a detailed view of how the renovations would impact the property.

The Challenge

This presented the sales/marketing team with a difficult challenge: How could they present what their property would look like once renovations were completed, in order to show potential customers the property so that pre-opening sales could begin taking place?

The Opportunity

The marketing goal is to create a real-life tour that would effectively reach potential customers, especially meeting professionals, to showcase the newly renovated convention hotel, and ultimately generate sales prior to its grand re-opening. The meeting space, hotel amenities, and general décor were quite unique; therefore the space needs to be carefully illustrated in order to convey these details to meeting professionals. Event diagram software with a feature of 3D rendering can depict a highly detailed presentation of what the renovated hotel would look like upon when finished.

How This Case May Be Applied to Advance Your Future Career

Many students in the event and convention management field will be involved in the sales of venues and services. The key message of this case study is to show the application of technology-enhanced venue diagrams and virtual tours for improved sales efforts. This case clearly represents exactly what the convention hotel would look like when finished so that pre-opening sales could take place. Understanding and gaining a skill set to apply an event diagram and virtual tour design tool can provide extensive background in the hospitality industry to effectively depict the experience of a meeting professional from start to finish. This skill set can differentiate you from others. More and more event venues adopt these programs to successfully bring their venue to market before it is actually open for use.

CASE #13: WHEN VM TECHNOLOGY ALLOWS YOU TO PLAN COST-EFFECTIVE AND HIGH-IMPACT SPEAKER SESSIONS

A meeting/event planner always wants to invite high-impact speakers to their event/meeting. However, most of time there are restrictions,- such as limited budget associated with travel costs or scheduling conflicts. Meeting technology, especially various levels of virtual meeting technologies, help planners to overcome those challenges to plan a high-impact speaker sessions.

The Challenge

Famous speakers attract attention and often allow meeting and event organizers to attract more people to their event. But in many occasions, the famous speakers are very busy, unable to travel, and their fees are high. Today's meeting planners should find an affordable but highly effective alternative speaker session. When organizing an event with a speaker who will participate via VMT, it is best to be prepared for the worse possible broadband connectivity. Broadband connection often varies depending on the speaker's residing country's technological infrastructure, and there is always the opportunity to have problems with the quality of the signal.

The Opportunity

Thanks to new technologies, meeting and event planners may invite famous speakers virtually and let them participate in events by using virtual meeting technology. This can reduce hosting organizers' expenses and allow them to stay within a budget while giving event planners an opportunity to have the speaker their audience wants. The speakers and audience who participate in VMT-embedded session should be educated in regard to tips on what to avoid (e.g. cloth, movement, eye contact, pace, etc.). Good news is that modern VMT is easy to apply and requires relatively minimal/manageable technology (hardware/skill set) than ever along with affordable cost.

How This Case May Be Applied to Advance Your Future Career

Good planning and good preparation are the secret ingredients to a successful event. Having a backup plan is always the best preparation. Murphy's Law says "what can go wrong, most likely will go wrong." It may be the best statement that describes meeting/event management. The only way to overcome Murphy's Law (event problem) is to be prepared and have contingency plans. Students or young meeting/event planners should understand that technology-embedded sessions can face problems such as connectivity problems. It is ideal to have technical staff ready to work on the problem. VMT-embedded sessions can be used in meetings and events as it is or as backup plan to live speaker sessions. VMT technology provides a recording feature so a recorded version of the speech can be viewed later. Further these recorded sessions can generate revenue by packaging as a continuing education model for those who couldn't attend the live session.

CASE #14: LIVE EVENT MOBILE ON-SITE ATTENDEE ENGAGEMENT

Event participants are increasingly engaging in second-screen devices while watching live events. A meeting/event venue starts with designing a Wi-Fi system capable of supporting event-day crowds and with integrating a second-screen app into the network. This can offer attendees live commentary, exclusive content, and social sharing features.

The Challenge

Large event venues (e.g. major convention centers, arenas, and stadiums) are some of the few remaining "black holes" in regard to Wi-Fi connectivity. At the same time, social media has become a key component of attendee engagement during live events at an explosive rate. Event attendees are eager for live updates, in-session commentary, and the ability to share their event experiences socially. Many event venues and organizers are expected to offer a solution to this infrastructure challenge and to develop an app to meet attendees' increasing demands for mobile engagement.

The Opportunity

Today's technology offers a solution for infrastructure, called a high-density Wi-Fi that can integrated with an event-day app smoothly. An event-day app can offer exclusive content and serve as a connecting point for attendees, organizers/sponsors, and meeting planners. Tailored with the attendees in mind, the app allows them to keep up to date on programs, follow live sessions, and vote for in-class. The app serves as a second-screen partner for each event's live action and makes it easy share content on Twitter and Facebook. To promote the app, consider a social media marketing strategy that encourages attendees to vote on content using Twitter hashtags.

How This Case May Be Applied to Advance Your Future Career

Many meeting and event venues are required to provide a high-density Wi-Fi system capable of servicing packed venues. It is critical to find a provider that is capable of providing reliable connectivity for a large crowd. Consider how to develop a mobile solution that is able to capture attendee data. An ability to develop an app that is able to deliver quality content and enhance event experience is in big demand from the event and meeting industry. Reacting to events in real time, you need to understand how to keep the content from apps updated in order to engage attendees at all points throughout an event and maintain the conversation going, even after the event ended.

Index

An environmentally friendly book printed and bound in England by www.printondemand-worldwide.com

PEFC Certified

This product is
from sustainably
managed forests
and controlled
sources

PEFC/16-33-415

www.pefc.org

This book is made of chain-of-custody materials; FSC materials for the cover and PEFC materials for the text pages.